Evergreen Oak Woodlands' Role in Tackling Climate Change and Preserving Mediterranean Landscapes

Evergreen Oak Woodlands' Role in Tackling Climate Change and Preserving Mediterranean Landscapes

Edited by

Rocco Carella

Cambridge Scholars Publishing

Evergreen Oak Woodlands' Role in Tackling Climate Change and Preserving Mediterranean Landscapes

Edited by Rocco Carella

This book first published 2023

Cambridge Scholars Publishing

Lady Stephenson Library, Newcastle upon Tyne, NE6 2PA, UK

British Library Cataloguing in Publication Data
A catalogue record for this book is available from the British Library

ISBN (10): 1-5275-2916-9
ISBN (13): 978-1-5275-2916-8

CONTENTS

CONTRIBUTORS

Rocco Carella (Editor), *Dr. in Forest Science, Ph.D. in Landscape Analysis, Environmental Consultant. See:*
https://www.studioambientale-forestaleroccocarella.it/

Roberto Mercurio, *former Full Professor of Siliviculture and Forest Management, Mediterranean University, Reggio Calabria, Italy.*

Roubina Ghattas, *M.Sc. in Utilization and Conservation of Plant Genetic Resources – Birmingham University UK, Head of Biodiversity and Food Security Department – The Applied Research Institute-Jerusalem (ARIJ).*

Munir Ozturk, *Prof. Dr., Centre for Environmental Studies/Botany Dept., Ege University, Izmir, Turkey & Universiti Putra Malaysia, Selangor, Malaysia.*

Volkan Altay, *Prof. Dr., Centre for Environmental Studies/Botany Dept., Ege University, Izmir, Turkey & Universiti Putra Malaysia, Selangor, Malaysia.*

CHAPTER 1

INTRODUCTION

ROCCO CARELLA

The Mediterranean basin is one of the most important biodiversity hotspots on Earth, but at the same time one of the regions most affected by the dramatic consequences of climate crisis. As is well known, one of the most typical aspects of spontaneous vegetation in the Mediterranean basin are the evergreen coenoses in their different habitus, often the result of anthropogenic disturbance. In the forest environment, an outstanding and peculiar element is the well-known diverse complex of numerous sclerophyllous species, where a fundamental role is played by evergreen oaks. This text provides contributions that aim at improving and underlining the knowledge of three evergreen Mediterranean oaks – the Palestine oak, the holly oak and the cork oak – and their indispensable function in biodiversity and landscape conservation, which is more important than ever in the challenging global environment.

CHAPTER 2

A CONTRIBUTION TO THE ECOLOGICAL UNDERSTANDING OF A LESSER-KNOWN MEDITERRANEAN FOREST SPECIES. THE STATUS AND ECOLOGY OF THE PALESTINE OAK (*QUERCUS CALLIPRINOS* WEBB) IN THE MURGIA PLATEAU (APULIA REGION, SOUTH ITALY)

ROCCO CARELLA

2.1 Introduction

The Palestine oak (*Quercus calliprinos* Webb) is of great phytogeographical interest in Italy. For it is here that the most western populations of its natural distribution occur (Fig. 2–1).

Quercus calliprinos Webb is related to *Quercus coccifera* L., but differs in its geographical distribution and morphological aspects – such as the young yellowish twigs, stellate pubescent persisting until the second year and the female catkins tomentose (Pignatti, 2002). Some authors consider *Quercus calliprinos* as a subspecies of *Quercus coccifera* (Tutin et al., 1993) or a morphotype (Toumi & Loumaret, 2010), others as a valid species (Camus, 1936–38; Zohari, 1961; Pignatti, 2002).

While the holly oak (*Quercus coccifera)* is a shrub that lives in the western districts of the Mediterranean basin, the Palestine oak is a small tree up to 10 metres in height and distributed in the eastern part of the Mediterranean basin (Italy, Greece, Albania, the former Yugoslavia, Turkey, Lebanon, Syria, Palestine, Israel, Jordan and Saudi Arabia, as well in the north-western Hijaz mountains according to Ghazanfar & Fischer (1998) (Fig. 2–1), with a different ecology since the species lives from the coastal line up to mountain level (Serruya, 1978).

Fig. 2–1. Natural distribution of Quercus calliprinos *and* Quercus coccifera *(Gianguzzi L. & La Mantia A., Azienda Foreste Demaniali-Regione Sicilia, 2000).*

According to some authors, in some districts of the natural distribution both *Quercus coccifera* and *Quercus calliprinos* can live (Zohari & Orshan, 1965; Gentile & Gastaldo, 1976).

Fig. 2–2. Leaf details of Quercus calliprinos *(Photo Rocco Carella).*

Fig. 2–3. Details of the lower side of the leaves of Quercus calliprinos *(Photo Rocco Carella).*

Fig. 2–4. Male inflorescence of Quercus calliprinos *(Photo Rocco Carella).*

Fig. 2–5. A flowering individual of Quercus calliprinos *(Photo Rocco Carella).*

Fig. 2–6. Details of Quercus calliprinos *acorns (Photo Rocco Carella).*

Fig. 2–7. Immature acorn (Photo Rocco Carella).

In Italy, spontaneous *Quercus calliprinos* coenoses occur in Apulia, in Basilicata –just in the Senise area in the southern part of Potenza province (Gavioli, 1935) – and on the two main islands, Sicily and Sardinia (Paffetti et al., 2001; La Mantia & Gianguzzi, 2003).

The Apulian natural distribution of *Quercus calliprinos* is discontinuous, with two distinct areas within the regional territory (Fig. 2–8). For some authors, the anthropogenic impact and, in particular, agricultural practices (Francini-Corti, 1966) are the reason for this peculiar distribution, leading to the deep alteration of the original vegetation mosaic (Chiesura Lorenzoni, 1974).

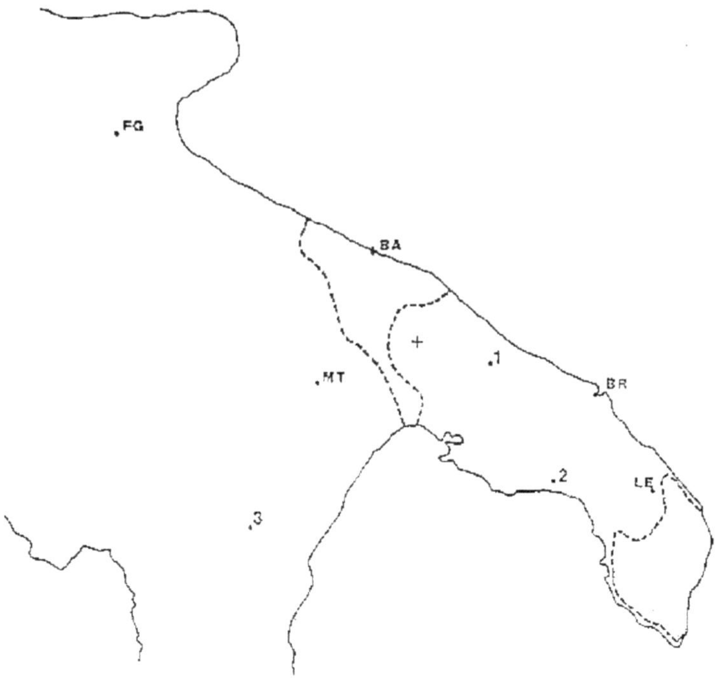

Fig. 2–8. Apulian natural distribution of Quercus calliprinos *(in Bianco* et al.*, 1989).*

The southern Apulian populations that are located in southern Salento, and the northern Apulian populations of the Bari province, differ in ecological considerations, since in Salento *Quercus calliprinos* grows in termophilous maquis, and also on cliffs along the sea (Fig. 2–9), where the species can often be seen growing alongside the holm oak (*Quercus ilex L.*) (Sabato, 1972; Lorenzoni & Chiesura-Lorenzoni, 1987) and other sclerophyllous scrubs. In the northern part of the Apulian region, on the other hand, the species forms maquis and open forests that are spread mostly across that peculiar geographical district known as Conca di Bari (Bari basin) (Baldacci, 1962) (Fig. 2–10) or its borders, in the Murgia escarpment around Cassano delle Murge and, more rarely, in the High Murgia Plateau. Differently from what was recorded in Salento, the Palestine oak populations of the northern part of the Apulian distribution are not observed along the coastal line, and can also live in hilly districts, several kilometres away from the Adriatic sea.

Another important aspect is that in the Bari province *Quercus calliprinos* populations do not form mixed coenoses with *Quercus ilex*. The holm oak is very rare in this part of the Murgia Plateau and occurs locally only where the disappearance of Palestine oak is recorded, as described for Lama Giotta (Carella, 2012) (Fig. 2–10), the southern ephemeral stream of the Bari basin.

Fig. 2–9. A stand of maquis with Quercus calliprinos *in the southern Salento on the Otranto Cape – Baia d'Orte (Photo Rocco Carella).*

Fig. 2–10. Holm oak high maquis along the bank of Lama Giotta ephemeral stream (Photo Rocco Carella).

The most important spontaneous coenoses with Palestine oak of the Murgia Plateau in Bari province have been investigated, leading to the results here presented. Nevertheless, in works regarding the study area, some authors have considered the species as *Quercus coccifera s.l.* or *Quercus coccifera* (Vita et al., 2001); the coenoses described in the present research are referred to as Palestine oak (*Quercus calliprinos*), which is considered a valid species. In the study area it is possible to find stands of Palestine oak with habitus of low maquis, but it is important to underline that whenever coenoses are not touched by fire for years, the physiognomy turns to high maquis forest; so locally solitary small trees of veteran individual *Quercus calliprinos* and relic undisturbed patches of high maquis and maquis forest up to 8 metres in height are observed (Fig. 2–11).

Fig. 2–11. Solitary veteran individual of Quercus calliprinos *(Acquaviva delle Fonti – Masseria La Rena, Photo Rocco Carella).*

2.2 Description of the study area

2.2.1 Geo-morphological and climatic aspects

The Murgia Plateau is a cretaceous limestone plateau of the Apulian platform, and here the soils are in general superficial *Terre Rosse*, while along the ephemeral streams deeper soils on the Quartenary alluvial sediments can exclusively be observed. In some portions of the plateau, on the cretaceous limestone, Pleistocenic deposits, known as *Tufi delle Murge*, can locally be observed (Azzaroli & Valduga, 1967).

Karsism is very evident in the whole plateau, with different phenomena where the most famous and amazing episodes are represented by deep caves, such as the Castellana Caves, and huge dolines, such as Pulo di Altamura (Fig. 2–12). As a consequence of its geomorphology, the Murgia Plateau shows a lack in the superficial waters.

Fig. 2–12. A view of Pulo di Altamura, the hugest doline of the Murgia Plateau (Photo Rocco Carella).

According to Rivas Martinez (2004), the whole Murgia Plateau occurs in the Mediterranean region, the Mediterranean Oriental sub-region, the Adriatica province, the Apuliana sub-province; the bioclimatic belt is meso-Mediterranean and the bioclimate is Mediterranean pluviseasonal oceanic.

The altitude of the Palestine oak stands of the Bari basin can vary from 40 to 355 metres above sea level. The following table shows the main climatic parameters recorded at Casamassima in the central part of the Bari basin (Tab. 2–1).

Yearly average temperature	**14.7°C**
Wearly average precipitation	**545.8 mm**
Average temperature of the coldest month of the year	**7.1°C**
Average temperature of the warmest month of the year	**23.4°C**
Average temperature of the absolute minimum	**-2.9°C**
Average temperature minimum of the coldest month of the year	**3.5 °C**
Average temperature maximum of the coldest month of the year	**10.7°C**
Continentality Index (Rivas Martinez)	**Euoceanic**
Termicity Index (Rivas Martinez)	**290**
Ombrotype (Rivas Martinez)	**Upper dry (Udry)**
Bioclimatic belt (Rivas Martinez)	**Transition between Lower mesomediterranean (Lmme) and Upper mesomediterranean (Umme)**
Winter Cold Stress Index (WCS Mitrakos)	**145.6**
Year Cold Stress Index (YCS Mitrakos)	**180.8**
Summer Drought Stress Index (SDS Mitrakos)	**141.2**
Year Drought Stress Index (YDS Mitrakos)	**206.2**
Yearly Evapotraspiration ETP according Thornthwaite and Mather (1955)	**785.4 mm**

Tab. 2–1. Main climate parameters of Casamassima (223 metres above sea level, period of observation 1967–96).

In the following figure is shown the evapotranspiration diagram for Casamassima (period of observation 1967–96) according to Thornthwaite & Mather (1955).

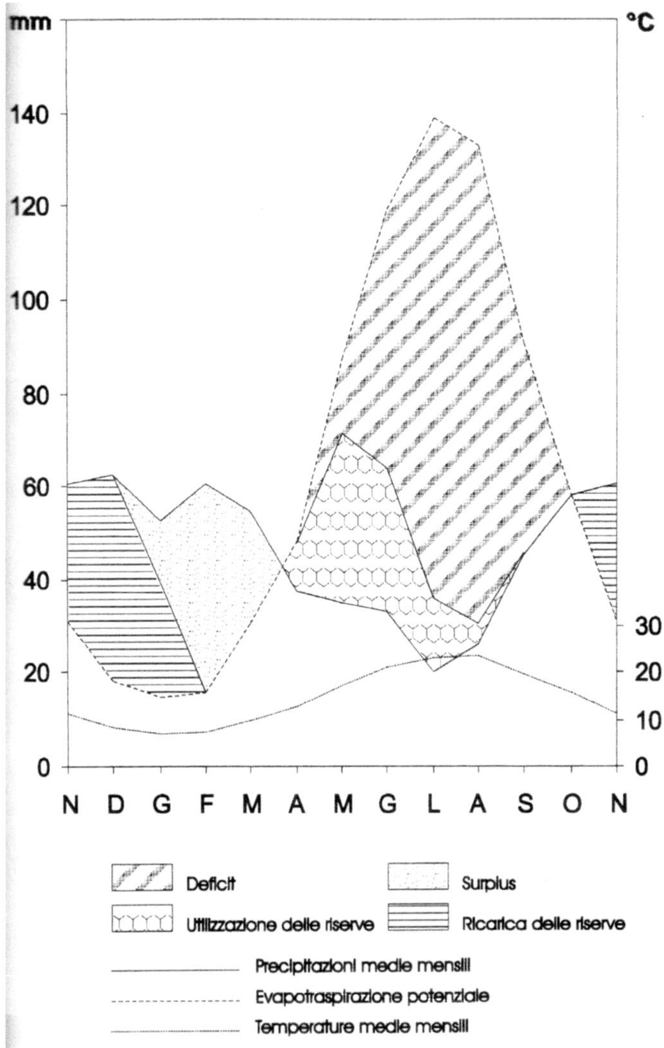

Fig. 2–13. ETP diagram of Casamassima (223 metres above sea level, period of observation 1967–96).

Table 2–2 shows the main climatic parameters recorded at Cassano delle Murge, where the Palestine oak is abundant on the Murge escarpment at the margin of the Bari basin.

Yearly average temperature	14.7°C
Wearly average precipitation	638.7 mm
Average temperature of the coldest month of the year	6.8°C
Average temperature of the warmest month of the year	24.0°C
Average temperature of the absolute minimum	-2.7°C
Average temperature minimum of the coldest month of the year	3.5 °C
Average temperature maximum of the coldest month of the year	9.5°C
Continentality Index (Rivas Martinez)	Euoceanic
Termicity Index (Rivas Martinez)	277
Ombrotype (Rivas Martinez)	Upper dry (Udry)
Bioclimatic belt (Rivas Martinez)	Upper mesomediterranean (Umme)
Winter Cold Stress (WCS) Index (Mitrakos)	168.3
Year Cold Stress (YCS) Index (Mitrakos)	196.3
Summer Drought Stress (SDS) Index (Mitrakos)	86.8
Year Drought Stress (YDS) Index (Mitrakos)	118.2
Yearly Evapotraspiration ETP according Thornthwaite and Mather (1955)	791.0 mm

Tab. 2–2. Main climate and bioclimatic parameters of Cassano delle Murge (420 metres above sea level, years of observation 1967–96).

The following figure (Fig. 2–14) is the Montero de Burgos bioclimatic diagram (1973) of Cassano delle Murge.

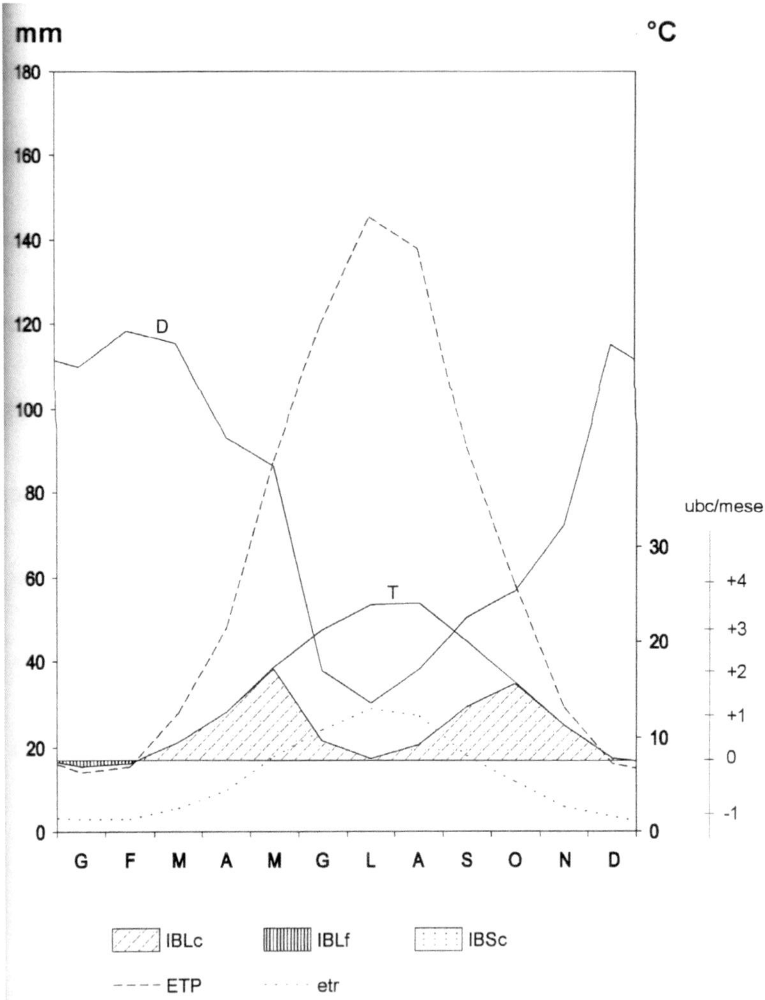

Fig. 2–14. Montero de Burgos bioclimatic diagram. Cassano delle Murge (420 metres above sea level, years of observation 1967–96).

2.2.2 Landscape and distribution of Palestine oak coenoses

The northern part of the Apulian distribution of the Palestine oak is spread in particular in the Bari province. Nevertheless, some authors, such as Giannuoli (1951), Crivellari (1950), Bianco et al. (1981–82), have in the past also described some stands in the south-western district near Mottola in the province of Taranto.

The following map (Fig. 2–15) shows the distribution of the *Quercus calliprinos* dominated coenoses in the Bari province.

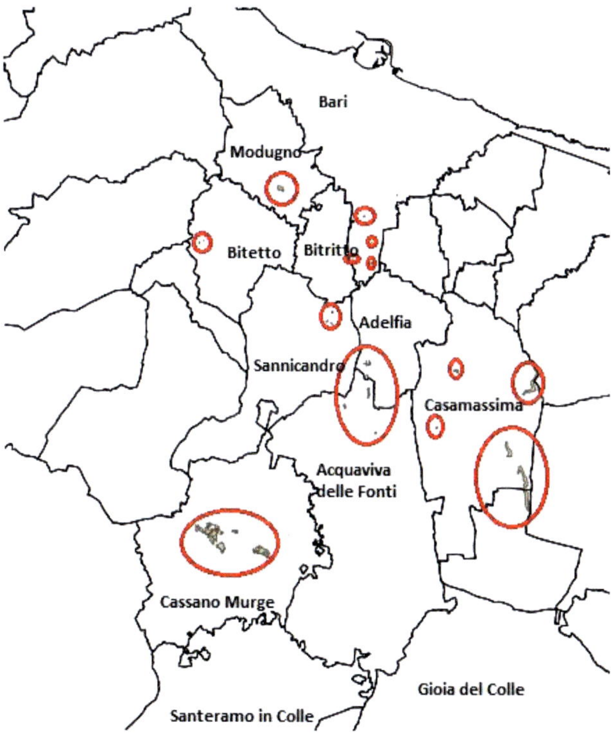

Fig. 2–15. Distribution of the Quercus calliprinos *dominated coenoses in the Bari province.*

The Palestine oak stands of the Bari basin district occur in particular along slopes and rocky banks of the ephemeral streams (Figs. 2–16, 2–17).

Fig. 2–16. Patches of Quercus calliprinos *maquis on the rocky bank of Torrente Baronale ephemeral stream (Adelfia, Bari province) (Photo Rocco Carella).*

Fig. 2–17 Quercus calliprinos *maquis on the rocky bank of the lower course of Baronale ephemeral stream near Loseto (Photo Rocco Carella).*

These streams, familiarly known as *lame*, are a peculiar landscape feature of the Bari basin (Carella, 2015), where they start in the High Murgia and cut through the Low Murgia before reaching the Adriatic sea in the Bari area. This confluence of several erosion streams in the hinterland of Bari gives the name Conca di Bari to this part of the low Murgia Plateau, for its similarity with a basin. Palestine oak occurs in some of these ephemeral streams and, in particular, moving in a north–south direction, *Quercus calliprinos* is observed along Lama Lamasinata, Torrente Picone (Carella, 2013a) and its two branches Torrente Baronale and Torrente La Badessa, Lama La Fitta (Carella, 2017; Carella, 2013b), Lama Cutizza (Carella, 2013c), and Lama San Giorgio (Carella et al., 2003).

Stands of Palestine oak of the Bari basin are particularly abundant along Torrente Picone and its two branches, where several patches of maquis of Palestine oak live along all its course. It is here that some of the most representative *Quercus calliprinos* high maquis stands of the whole

Murgia Plateau can be appreciated. Rarely can small relic coenoses be observed in the Bari basin as well as outside the ephemeral stream, as happens in the area between Casamassima and Acquaviva delle Fonti.

In the Bari basin, furthermore, often *Quercus calliprinos* can be observed along the dry stone walls in the olive fields that are typical for this territory and along the rural roads (Fig. 2–18), where sometimes there occur dense hedges of Palestine oak up to 4–6 metres in height (Fig. 2–19).

Fig. 2–18. High maquis of Quercus calliprinos *along a rural road in the Bari basin, Adelfia (Photo Rocco Carella).*

Fig. 2–19. Dense high hedge (up to 5 metres) of Quercus calliprinos *within olive fields in the Bari basin, Bitritto (Photo Rocco Carella).*

Quercus calliprinos also lives in the Murgia escarpment around Cassano delle Murge, which represents the border of the Bari basin, and in other scattered stands outside the Bari basin, in the High Murgia. In the Murgia escarpment near Cassano delle Murge *Quercus calliprinos* is abundant (Fig. 2–20) at altitudes between 370 and 430 metres above sea level. Here the species forms high maquis (low maquis occurs only where fire has recently passed) (Fig. 2–21), open forests (Figs. 2–22, 2–23, 2–24) and can also enter mixed deciduous–evergreen forests, with the Palestine oak in the dominated layer and the Italian pubescent oak (*Quercus virgiliana*) in the dominant layer,[1] as observed near Grotta del Lupo (Armenise, 1952) (Fig. 2–25).

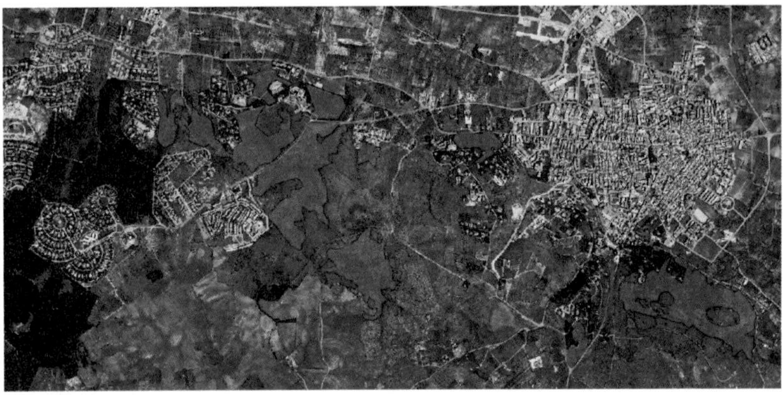

Fig. 2–20. Distribution of Quercus calliprinos *dominated coenoses on the escarpment near Cassano delle Murge.*

[1] *Quercus dalechampii* can locally appear in the dominant layer.

Fig. 2–21. Low maquis of Quercus calliprinos *a few years after fire. Abbey of Agostiniani, Cassano delle Murge (Photo Rocco Carella).*

Fig. 2–22. Open forest of Quercus calliprinos *in the escarpment near Cassano delle Murge at Serra di Laudati (Photo Rocco Carella).*

Fig. 2–23. Open forest of Quercus calliprinos *in the escarpment near Cassano delle Murge (Photo Rocco Carella).*

Fig. 2–24. Scrub presence of Quercus calliprinos *in forest patches dominated by* Quercus virgiliana *near Cassano delle Murge (Photo Rocco Carella).*

Fig. 2–25. Mixed forest with Quercus calliprinos *and* Quercus virgiliana *at Grotta del Lupo, Cassano delle Murge (Photo Rocco Carella).*

A few stands with *Quercus calliprinos* are also observed in the High Murge Plateau far from the Cassano delle Murge escarpment, outside the Bari basin, at altitudes between 400 and 495 metres above sea level. In these stands the species mostly enters locally in the scrub layer of deciduous termophilous oak-dominated forests, as observed sometimes in the Mesola area (Carella, 2014) (Figs. 2–26, 2–27), Bosco della Parata, Bosco Il Quarto (Vita et al., 2002). Similar coenoses are very rare in the Bari basin, but have been recorded in particular at Bosco di Grottagiglio (Fig. 2–28).

A special mention needs to be given to a small residual patch of open forests of *Quercus trojana* with *Quercus calliprinos* in the scrub layer, near Jazzo del Corvo, Altamura (Figs. 2–29, 2–30, 2–31), where there is a deeper penetration in the High Murgia towards the north-west not only of Palestine oak, but also Macedonian oak (Carella, 2011).

Fig. 2–26. Mixed forest with Quercus calliprinos, Quercus virgiliana *and* Quercus trojana *at Mesola Forest, Cassano delle Murge (Photo Rocco Carella).*

Fig. 2–27. A further view of the Mesola Forest (Photo Rocco Carella).

Fig. 2–28. Mixed forest with Quercus virgiliana *and* Quercus calliprinos *at Grottagiglio Wood, Cassano delle Murge (Photo Rocco Carella).*

Fig. 2–29. The location of the stand of the residual patch of open forest of Macedonian oak with Quercus calliprinos *at Jazzo del Corvo, Altamura.*

Fig. 2–30. The residual open forest of Macedonian oak with Quercus calliprinos *at Jazzo del Corvo, Altamura (Photo Rocco Carella).*

Fig. 2–31. A further view of the relic patch with Macedonian oak and Palestine oak scrubs at Jazzo del Corvo, Altamura (Photo Rocco Carella).

2.2.3 Vegetation aspects

The Murge Plateau is generally distinguished by two landscape units: the north-western sector and the south-eastern sector. The northern-western district covers a territory spread across the new BAT (Barletta-Andria-Trani) province and in the Bari province, and is characterized by Mediterranean pastures (known as Mediterranean pseudo-steppes) (Figs. 2–32, 2–33) – a habitat that is of great relevance to biodiversity conservation (European Commission, 2013) for its floristic, vegetation and faunistical richness. Depending on the composition and evolution level, these herbaceous coenoses can be given the habitat code 6220* *Pseudo-steppes with grasses and annuals of the* Thero-Brachypodietea, or code 6210* *Semi-natural dry grassland and scrubland facies on calcareous substrates* (Festuco-Brometalia) *(Important orchid sites)* of Annex I of the 92/43/EEC Directive. Recently the pastures of the High Murge Plateau have been described as the association *Acinos suaveolentis-Stipetum austroitalicae* (Forte et al., 2005) included in the order *Scorzonero-Crysopogonetalia* Horvatic & Horvat, 1956 1958, of the alliance *Hippocrepido glauce-Stipion austroitalicae* (Forte & Terzi, 2005) of the class *Festuco-Brometea* Br.-Bl and Tx-, 1942, ex Klika & Hadac, 1944; this vegetation type is given the code 62A0 *Eastern sub-mediterranean dry grasslands* (*Scorzoneretalia villosae*) in Annex 1.

Fig. 2–32. Mediterranean pseudo-steppes in the northern-western sector of Murge Plateau (Photo Rocco Carella).

Fig. 2–33. Pseudo-steppes and artificial conifer forest stands, the typical landscape of the northern-western Murgia Plateau (Photo Rocco Carella).

Forests here are mostly represented by spontaneous deciduous termophilous coenoses dominated by *Quercus virgiliana* (Ten.) Ten. (Fig. 2–34) and artificial forest stands dominated by *Pinus halepensis* Miller and other coniferous species established to prevent erosion mostly along slopes (Fig. 2–35).

Biondi et al. (2004) have described for the deciduous termophilous coenoses a new association *Stipo bromoidis-Quercetum dalechampii,* included in the alliance *Carpinion orientalis* Horvat 1958, of the order *Quercetalia pubescenti-peraeae* Kilka 1933 of the classs *Querco-Fagetea* Br.-Bl &_Vlieger in Vlieger 1937. These forest communities have recently mostly been given the code 91AA* *Eastern white oak woods* of Annex I of the Habitat Directive.

Fig- 2–34. Forest coenoses dominated by Quercus virgiliana *in the High Murgia Plateau (Bosco il Quarto) (Photo Rocco Carella).*

Fig. 2.35. Artificial forest stand of Pinus halepensis *in the north-western sector of Murgia Plateau (Photo Rocco Carella).*

The south-eastern sector of the Murgia Plateau starts in the Bari province in the valley of Gioia del Colle, which cuts through the two districts of the plateau, but is mostly in the Taranto province. The landscape mosaic in this district is typically characterized by wheat fields with scattered forest patches dominated by Macedonian oak (*Quercus trojana* Webb) (Fig. 2–36), a species of great phytogeograpical interest, since the Apulian populations of the southern-western district of Murgia that trespass the regional boundaries in woods around Matera (Gavioli, 1935) in the closed Basilicata region are the most western coenoses of the natural distribution of this Balkan species.

Macedonian oak is referred to as *Euphorbio apii-Quercetum trojanae* (Bianco et al., 1998) and is included in the alliance *Fraxino orni-Quercetum ilicis* (Biondi, Casavecchia & Gigante 2003), of the order *Quercetalia ilicis* (Br.-Bl. Ex Molinier, 1934), and of the class *Quercetea ilicis* (Br.Bl. Ex A. & O. Bolios 1950). While Biondi et al. (2004) have described a new association, *Teucrio siculi-Quercetum trojanae*, for the *Quercus trojana* coenoses of the summit areas of the gorges (*gravine*) (Fig. 2 - 37) of the southern border of the southern-eastern district of the Murgia Plateau.

The Macedonian oak coenoses are included in code 9250, Quercus trojana *woods* of Annex I of the Habitat Directive.

Fig. 2–36. Landscape of the southern-eastern sector of the Murgia Plateau with solitary individuals and forest patches of Quercus trojana *(Photo Rocco Carella).*

Fig. 2–37. A view of the Palagianello Gravina, *one of the most important and deep gorges of the south-eastern Murgia Plateau (Photo Rocco Carella).*

2.3 Vegetation analysis

2.3.1 Phytosociological relevés

Relevés according to the phytosociological method have been conducted in some of the most important spontaneous coenoses with Palestine oak of the study area.

The following map (Fig. 2–38) shows the location of the surveys, selected from the most representative coenoses of the study area. To collect evidence of the ecology of the Palestine oak in the Murgia of Bari, surveys have been conducted both in maquis of Palestine oak (relevés 1–6) and in deciduous termophylous stands of Italian pubescent oak, with *Quercus calliprinos* in the dominated/scrub layer (relevés 7–11).

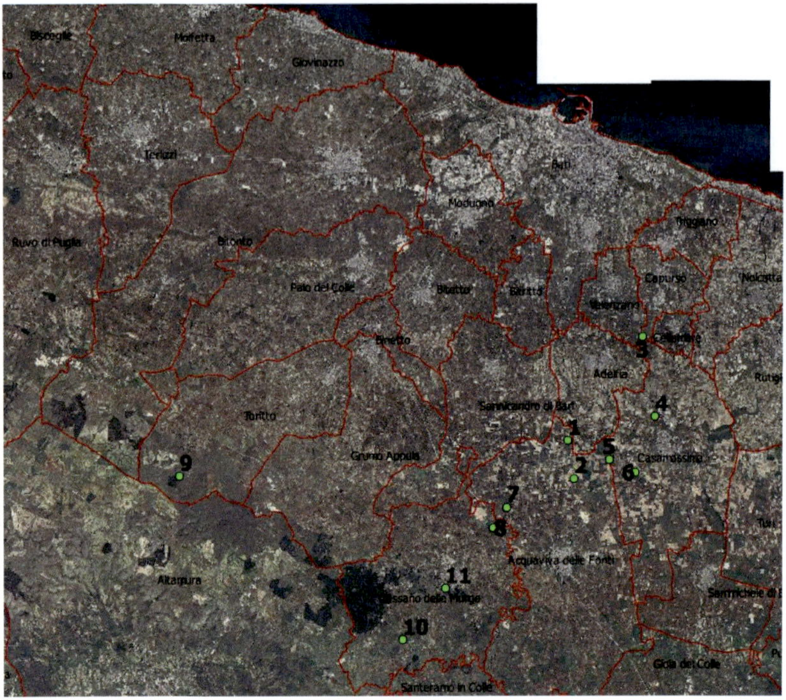

Fig. 2–38. Map showing the location of the phytosociological relevés.

Relevé n.1

Date: 8 April 2009
Author: Rocco Carella
Location: Lago dell'Arciprete (Baronale ephemeral stream)
Municipality: Acquaviva delle Fonti
Exposure: W N/W
Altitude: 205 m a.s.l.
Surface: 100 m^2
Inclination: 5%
Rock cover: 5%
Height of tree layer: 1.5–2 m
Cover of tree layer: 70%

Tree layer
Quercus calliprinos 4.4
Calicotome villosa 1.2
Pistacia lentiscus +

Scrub layer
Osyris alba +
Prasium majus +
Rosa canina +
Teucrium polium +

Herbaceous layer
Rubia peregrina +
Poa bulbosa +
Geranium robertianum +
Asphodelus microcarpus 1.2
Ornithogalum umbellatum +
Carex flacca +
Tetrogonobolus purpureus +

Notes: In the clearings of the maquis occurs *Stipa austroitalica* and numerous orchis species; particularly abundant is *Orchis papilionacea*.

Fig. 2–39. Maquis of Quercus calliprinos *at Lago dell'Arciprete, Acquaviva delle Fonti (Photo Rocco Carella).*

Relevé n.2

Date: 9 April 2009
Author: Rocco Carella
Location: Valzanello
Municipality: Acquaviva delle Fonti
Exposure: null
Altitude: 245 m a.s.l.
Surface: 90 m^2
Inclination: 5%
Rock cover: 5%
Height of tree layer: 2–4 m
Cover of tree layer: 80%

Tree layer
Quercus calliprinos 4.4
Quercus virgiliana 1.2
Pistacia terebinthus 1.2
Smilax aspera +
Hedera helix +

Scrub layer
Asparagus acutifolius +
Pistacia lentiscus +
Quercus virgiliana +
Pistacia terebinthus +
Prasium majus +

Herbaceous layer
Cyclamen hederifolium 1.2
Geranium robertianum 1.2
Hedera helix +
Acinos suaveolens +
Carex flacca +
Dasypyrum villosum +
Filipendula vulgaris+

Relevé n. 3

Date: 12 April 2009
Author: Rocco Carella
Location: *Pacifico*
Municipality: Capurso
Exposure: S-S/E
Altitude: 140 m a.s.l.
Surface: 80 m^2
Inclination: 5%
Rock cover: 5%
Height of tree layer: 2–4 m
Cover of tree layer: 80%

Tree layer
Quercus calliprinos 4.4
Olea europaea var. sylvestirs +
Calicotome villosa 1

Scrub layer
Calicotome villosa 2.2
Pistacia lentiscus +
Asparagus acutifolius +
Ficus carica +

Herbaceous layer
Rubia peregrina +
Tamus communis +
Trifolium stellatum +
Scorpiurus muricatus +
Vicia benghalensis +
Silene vulgaris +
Dactylis hispanica +

Notes: Arson occurs very frequently every summer, sometimes several times in the same season. In the clearings of the maquis live numerous orchid species; particularly abundant are *Orchis morio* and *Orchis papilionacea.*

Fig. 2–40. A patch of maquis of Quercus calliprinos *at Contrada Pacifico, Capurso (Photo Rocco Carella).*

Relevè n. 4

Date: 19 April 2009
Author: Rocco Carella
Location: Masseria Azzone
Municipality: Casamassima
Exposure: W N/W
Altitude: 207 m a.s.l.
Surface: 100 m^2
Inclination: 0 %
Rock cover: 5–10 %
Height of tree layer: 7–8 m
Cover of tree layer: 100%

Tree layer
Quercus calliprinos 5.4
Viburnum tinus 1.2
Rhamnus alaternus +

Scrub layer
Quercus virgiliana +
Crataegus monogyna +
Asparagus acutifolius +
Pistacia lentiscus +
Viburnum tinus +
Rhamnus alaternus +
Daphne gnidium +
Tamus communis +

Herbaceous layer
Buglossoides purpurocaerulea 1.2
Cyclamen hederifolium +
Geranium robertianum +
Rubia peregrina +

Relevé n. 5

Date: 19 April 2009
Author: Rocco Carella
Location: Villa Azzone
Municipality: Acquaviva delle Fonti
Altitude: 245 m a.s.l.
Surface: 100 m^2
Inclination: 0%
Rock surface: 0–5%
Height of tree layer: 7–8 m
Cover of tree layer: 70%

Tree layer
Quercus calliprinos 5.4

Quercus virgiliana +
Rhamnus alaternus +

Scrub layer
Osyris alba +
Rhamnus alaternus +
Phillyrea latifolia +
Pistacia lentiscus +
Lonicera etrusca +
Tamus communis +
Smilax aspera +

Herbaceous layer
Rubia peregrina 1
Allium subhirsutum +
Carex flacca +
Geranium robertianum +
Silene alba +
Melica transsylvanica +

Relevé n. 6

Date: 19 April 2009
Author: Rocco Carella
Location: Casino Pastore
Municipality: Casamassima
Altitude: 242 m a.s.l.
Surface: 100 m^2
Inclination: 0%
Rock cover: 0–5%
Height of tree layer: 8 m
Cover of tree layer: 100%

Tree layer
Quercus calliprinos 5.4
Quercus virgiliana +

Rhamnus alaternus +

Scrub layer
Rhamnus alaternus +
Phillyrea latifolia +
Pistacia lentiscus +
Lonicera etrusca +
Osyris alba +
Smilax aspera +
Tamus communis +

Herbaceous layer
Geranium robertianum 1.2
Rubia peregrina 1.2
Carex flacca +
Allium subhirsutum +
Silene vulgaris +
Melica transsylvanica +

Fig. 2–41. The relic high maquis of Quercus calliprinos *at Casino Pastore, Casamassima (Photo Rocco Carella).*

Relevè n. 7

Date: 27 April 2010
Author: Rocco Carella
Location: Bosco Perrillo
Municipality: Acquaviva delle Fonti
Altitude: 262 m a.s.l.
Surface: 150 m^2
Inclination: 0 %
Rock cover: 0%
Height of tree layer: 8–10 m
Cover of tree layer: 70–80%

Tree layer
Quercus virgiliana 4.4

Scrub layer
Quercus calliprinos 1.2
Crataegus monogyna 2.3
Rosa canina +
Rubus ulmifolius +
Prunus spinosa +
Calicotome villosa +
Cistus salvifolius 1

Herbaceous layer
Leopoldia comosa +
Asphodelus microcarpus +
Allium subhirsutum 1
Geranium robertianum 1
Asphodeline lutea 1
Geranium sanguineum +
Linum bienne +
Ferula communis +
Rubia peregrina +
Dactylis hispanica +

Allium roseum +
Vicia cracca +

Notes: A *Quercus virgiliana* coppice well started to conversion in high forest.

Relevé n. 8

Date: 26 April 2010
Author: Rocco Carella
Location: Bosco Grottagiglio
Municipality: Cassano delle Murge
Altitude: 285 m a.s.l.
Surface: 150 m^2
Inclination: 0%
Rock cover: 5%
Height of dominant tree layer (*Quercus virgiliana*): 8–10 m
Height of *Quercus calliprinos* layer: 2–4 m
Cover of dominant tree layer: 100%
Cover of dominated tree layer: 50–60%

Tree layer
Quercus calliprinos 3.1
Quercus virgiliana 4.4

Scrub layer
Crataegus monogyna 1
Rosa canina +
Asparagus acutifolius +
Osyris alba +

Herbaceous layer
Leopoldia comosa +
Urginea maritima +
Asphodelus microcarpus +
Pisum sativum +

Allium subhirsutum 2.3
Bromus hordeaceus +
Bromus erectus +
Paeonia mascula +
Geranium robertianum+
Geranium sanguineum +
Euphorbia apios +
Cyclamen hederifolium +
Buglossoides purpurocaerulea +
Geum urbanum +

Notes: *Quercus virgiliana coppice wi*th *Quercus calliprinos* in the dominated tree layer.

Fig. 2–42. Kermes vermilio *galls on a leaf of* Quercus calliprinos *observed at Bosco Grottagiglio, Cassano delle Murge (Photo Rocco Carella).*

Relevé n. 9

Date: 30 April 2010
Author: Rocco Carella
Location: Masseria La Sentinella – Jazzo Chieffi
Municipality: Toritto
Altitude: 465 m a.s.l.
Surface: 100 m^2
Inclination: 0%
Rock cover: 0%
Height of tree layer: 10 m
Cover of tree layer: 70%

Tree layer
Quercus virgiliana 4.4
Quercus frainetto +

Scrub layer
Viburnum tinum +
Quercus calliprinos +
Rubus ulmifolius 1
Osyris alba 1
Quercus virgiliana +
Rosa canina +
Crataegus monogyna +
Asparagus acutifolius +
Lonicera etrusca +
Prunus spinosa +

Herbaceous layer
Rubia peregrina 1
Asphodelus microcarpus +
Vicia cracca +
Hordeum bulbosum +
Clinipodium vulgare +
Buglossoides purpurocaerulea 1

Geranium robertianum +
Arum italicum +

Notes: Aged coppice. In the understory, outside the survey surface, *Paeonia mascula* has been observed.

Relevé n. 10

Date: 6 May 2010
Author: Rocco Carella
Location: Masseria Tarola
Municipality: Cassano delle Murge
Altitude: 450 m a.s.l.
Relevè area: 100 m^2
Inclination: 0%
Rock cover: 0%
Height of tree layer: 8–10 m
Cover of tree layer: 70%

Tree layer
Quercus virgiliana 4.4

Scrub layer
Quercus calliprinos +
Crataegus monogyna +
Asparagus acutifolius +
Lonicera etrusca +
Rubus ulmifolius +

Herbaceous layer
Geranium robertianum +
Ranunculus velutinus +
Asphodeline lutea +
Rubia peregrina +
Carex flacca+
Sanguisorba minor +

Teucrium chamaedrys +
Paeonia mascula +

Notes: *Quercus virgiliana* wood with *Quercus calliprinos* locally spread in the scrub layer and in the dominated tree layer, where *Mespilus germanica* can also be observed.

Relevé n. 11

Date: 13 May 2010
Author: Rocco Carella
Location: Parco S. Lucia
Municipality: Cassano delle Murge
Altitude: 410 m a.s.l.
Surface: 100 m^2
Inclination: 10 %
Exposure: ES
Rock cover: 5–10 %
Height of tree layer: 10–1 2 m
Cover of tree layer: 50–60 %

Tree layer
Quercus virgiliana 4.4

High scrub layer (dense up to 90% cover)
Quercus calliprinos 1.2
Phillyrea latifolia 2.3
Crataegus monogyna +
Tamus communis +
Smilax aspera +

Low scrub layer
Ruscus aculeatus 1
Asparagus acutifolius +
Osyrus alba +

Herbaceous layer
Buglossoides purpurocaerulea +
Carex distachya +
Ranunculus velutinus +
Rubia peregrina +
Asphodeline lutea +
Allium subhirsutum +
Bromus erectus +
Notes: *Anacamptis pyramidalis* is observed in the clearings. This is an interesting forest patch with a two-layer forest canopy.

2.3.2 Discussion

The Palestine oak is the main forest species of the Bari basin (Fig. 2–43) and is well spread, mostly in the portion of the High Murgia that borders the district being considered, and more rarely in scattered sites far from it.

The forest coenoses with Palestine oak occur in the study area mainly on slopes, on poor rocky soils. But sometimes it is also possible to appreciate them on deeper soils, as can happen in portions of the riverbeds of some ephemeral streams, in particular along Lama San Giorgio and Lama La Fitta.

The Palestine oak coenoses of the northern part of the Apulian distribution here described can be considered the most varied from an ecological point of view among the Italian populations of the species. Just in the Bari province, *Quercus calliprinos* also live in more mesophylous conditions, denoting an ecological distance with respect to the other Italian coenoses, which in general are more termophilous and are always referred to as the class *Quercetea ilicis*, sometimes as a degradation of the *Quercion ilicis*, other times to the *Pistacio-Rhamnetalia-alaterni,* for the abundance of species of *Oleo-Ceratonion,* or as well in a transition between *Oleo-Ceratonion* and *Quercion ilicis* (Chiesura Lorenzoni et al., 1974).

Fig. 2–43. The Bari basin district.

This aspect of degradation of the *Quercion ilicis*, which is typical for the *Quercus coccifera* populations of the western Mediterranean basin, the *Cocciferetum* of Provence (Braun-Blanquet, 1951) and the Palestine oak in Salento (*Arbuto-Quercetum calliprini; Hedero ilicis-Quercetum calliprini*), and in Sicily (*Junipero turbinatae-Quercetum calliprini; Chamaeropo humilis-Quercetum calliprinii*) (La Mantia & Gianguzzi, 2003) is not so evident in the Palestine oak coenoses of the Bari province, generally characterized by the scarce presence of the *Oleo-Ceratonion* species.

Among the associations reported above, Biondi et al. (op. cit.) have compared the Murgia of Bari Palestine oak populations to the *Hedero helicis-Quercetum calliprini,* which is also considered the climatophylous forest vegetation of the inland areas of Salento. If the considerations related to Salento can be accepted, the data from the new phytosociological surveys here reported show an ecological distance of the Murgia of Bari populations from the *Hedero helicis-Quercetum calliprini.*

The main differences between the vegetation of *Hedero helicis-Quercetum calliprini* and what is expressed in the 11 new surveys presented here are the total absence of some of the characteristics and differential species of the association, such as *Quercus ilex*, *Stipa bromoides*, *Iris collina* and *Brachypodium sylvaticum*, and the rare presence of others, such as *Ruscus aculeatus* and *Hedera helix.* Another important differential feature is the presence in the new surveys here reported of numerous mesophilous species of *Querco-Fagetea*, such as *Buglossoides purpurocaerulea*, *Rosa canina*, *Quercus virgiliana*, *Tamus communis*, *Lonicera etrusca*, *Crataegus monogyna*, *Geum urbanum*, *Clinopodium vulgare*, *Filipendula vulgaris* and *Quercus frainetto,* counterposed to the presence of a few species of *Oleo-Ceratonion*, such as *Prasium majus*, *Rhamnus alaternus*, *Olea europaea* var. *sylvestrys*, *Daphne gnidium* and *Pistacia lentiscus.*

The dominance of Palestine oak in the study area, the disappearance of holm oak, the fact that *Quercus calliprinos* enters different forest types forming both maquis dominated, but as well taking part of the deciduous termophilous forests typical of the High Murgia Plateau, and lastly the fact that its stands can be locally recorded very far from the coastal line and at altitudes around 400–450 metres above sea levels, suggest that, just in this part of Italy, *Quercus calliprinos* is able to show part of the full ecological behaviour that the species expresses in the core area of its natural distribution.

All these considerations, and the difficulty of reporting the Palestine oak coenoses here considered in the *Quercion ilicis*, open up the possibility of debate, as already partially exposed many years ago by Gentile (1962), on the presence of a different syntaxonomic scheme for the Palestine oak in the Italian Peninsula, also considering the ecological behaviour of the species in the Eastern Mediterranean basin (Nahal, 1961).

In the following tables the existing associations and syntaxonomical schemes in Italy (Tab. 2–3) and in the Eastern Mediterranean basin are listed (Tab. 2–4).

Association	Syntaxonomical scheme	Distribution	Characteristic species
Rusco aculeati-Quercetum calliprini **Mossa, 1990**	• Class ***Quercetea ilicis*** • Order ***Pistacio lentisci-Rhamnetalia alaterni*** • Alliance ***Juniperion turbinatae***	**Sardinia**	• ***Quercus calliprinos*** • ***Juniperus turbinata*** • ***Ruscus aculeatus*** • ***Phillyrea angustifolia***
Junipero turbinatae-Quercetum calliprini **Bartolo, Brullo & Marcenò, 1983**	• Class ***Quercetea ilicis*** • Order ***Pistacio lentisci-Rhamnetalia alaterni*** • Alliance ***Juniperion turbinatae***	**Southern Sicily**	• ***Quercus calliprinos*** • ***Juniperus turbinata***
Chamaeropo humilis-Quercetum calliprini **Brullo et Marcenò, 1985**	• Class ***Quercetea ilicis*** • Order ***Pistacio lentisci-Rhamnetalia alaterni*** • Alliance ***Oleo-Ceratonion siliquae***	**Western Sicily**	• ***Quercus calliprinos*** • ***Chamaerops humilis*** • ***Galium litorale***
Arbuto unedi – Quercetum calliprini **Brullo et al., 1986**	• Class ***Quercetea ilicis*** • Order ***Pistacio lentisci-Rhamnetalia alaterni*** • Alliance ***Oleo-Ceratonion siliquae***	**Sicily and Apulia (southern Salento)**	• ***Quercus calliprinos*** • ***Arbutus unedo***

Hedero helicis-Quercetum calliprini **Biondi et al., 2004**	• Class ***Quercetea ilicis*** • Order ***Quercetalia ilicis*** • Alliance ***Fraxino orni-Quercion ilicis*** • Sub-alliance ***Fraxino orni-Quercenion ilicis***	**Apulia (Salento and Murge)**	• ***Quercus calliprinos*** • ***Quercus ilex*** • ***Ruscus aculeatus*** • ***Stipa bromoides*** • ***Brachypodium sylvaticum*** • ***Hedera helix*** • ***Allium subhirsutum*** • ***Viburnum tinus*** • ***Iris collina***

Tab. 2–3. Italian Quercus calliprinos phytosociological associations.

Association	**Syntaxonomical scheme**	**Distribution**	**Characteristic species**
Pistacio palestinae-Quercetum calliprini **Zohari, 1960, Nahal, 1961**	• Class ***Quercetea calliprini-Quercetea ilicis*** • Order ***Quercetalia calliprini-Quercetalia ilicis*** • Alliance ***Quercion calliprini***	**Palestine, Lebanon, Syria, Turkey**	• ***Quercus calliprinos*** • ***Pistacia palaestina***
Crataegus azarolus-Quercetum calliprini **Zohari, 1972, Chikhali, 2000**	• Class ***Quercetea calliprini-Quercetea ilicis*** • Order ***Quercetalia calliprini-Quercetalia ilicis*** • Alliance ***Quercion calliprini***	**Palestine, Syria**	• ***Quercus calliprinos*** • ***Crataegus azarolus*** • ***Amygdalus korschinskii*** • ***Pyrus siriaca*** • ***Pistacia atlantica***
Pruno tortuosa-Quercetum calliprini	• Class ***Quercetea calliprini-Quercetea ilicis*** • Order ***Quercetalia calliprini-Quercetalia ilicis*** • Alliance ***Quercion calliprini***	**Palestine, Lebanon, Syria**	• ***Quercus calliprinos*** • ***Prunus tortuosa*** • ***Amygdalus orientalis*** • ***Poterium spinosum*** • ***Amygdalus spartioides*** • ***Crataegus monogyna*** • ***Acer monspessulanum***

Querco infectoria-Quercetum calliprini	• Class ***Quercetea calliprini-Quercetea ilicis*** • Order ***Quercetalia calliprini-Quercetalia ilicis*** • Alliance ***Quercion calliprini***	**Lebanon, Syria, Palestine, Jordan, Turkey**	• ***Quercus calliprinos*** • ***Quercus infectoria*** • ***Ruscus aculeatus*** • ***Rubia aucheri*** • ***Eryngium falcatum*** • ***Juniperus oxycedrus*** • ***Aristolochia altissima*** • ***Laurus nobilis*** • ***Myrtus communis***
Querco calliprinos-Phillyreetu m mediae **Martini, 1999**	• **Class *Quercetea calliprini-Quercetea ilicis*** • **Order *Quercetalia calliprini-Quercetalia ilicis*** • **Alliance *Quercion calliprini***	**Syria**	• ***Quercus callirpinos*** • ***Styrax officinalis*** • ***Rhamnus palaestina*** • ***Jasmiunium fruticans*** • ***Prunus ursina*** • ***Cistus villosus*** • ***Daphne oleifolia***
Styraco officinali-Quercetum calliprini	• Class ***Quercetea calliprini-Quercetea ilicis*** • Order ***Quercetalia calliprini-Quercetalia ilicis*** • Alliance ***Quercion calliprini***	**Syria**	

Tab. 2–4. Quercus calliprinos Eastern Mediterranean phytosociological associations.

2.3.3 Conclusion

The Palestine oak is the main forest species of the Bari basin and can also be found in the High Murgia, mostly in those areas in geographical continuity, as happens mainly in the escarpment around Cassano delle Murge and more rarely in stands far from here (Santeramo in Colle, Altamura, Toritto). In the study area *Quercus calliprinos* is, as explained, remarkable for its ecology, but also for its value to biodiversity and landscape conservation; this is particularly evident in the Bari basin, where the Palestine oak high maquis along the ephemeral streams are a typical feature of this territory and the *Quercus calliprinos* coenoses also improve the ecological connectivity of these corridors (MacDonald, 2003), providing others important ecosystem services. Furthermore, the Palestine oak, as already observed by other authors (Chiesura-Lorenzoni & Lorenzoni, 1979), in the study area also has a strong regenerative capability after fire.

These precious habitats suffer in particular in the Bari basin from the lack of institutional protection since these coenoses are not included in parks, or Natura 2000 sites. Another serious gap is that, at the moment, the Palestine oak maquis forests are not listed in the regional territory as a habitat of Annex I 92/43/EEC Directive, where they should be designated code 93A0, *Woodland with* Quercus infectoria (Anagyro foetidae-Quercetum infectoriae) (European Commission, 2013). The Bari basin status of Palestine oak coenoses is even more complicated, since the considered coenoses are here exposed to different types of anthropogenic disturbances.

The importance of some of the coenoses here described also relates to their peculiar canopy structure, as happens in those stands where *Quercus virgiliana*, or *Quercus trojana*, or both, form the dominant layer, and the Palestine oak the dominated. These peculiar forest patches should be considered as areas of great naturalistic value, and adequate silvicultural treatments are needed to preserve the compositions and structure of these forest stands. These mixed deciduous–evergreen coenoses are also important to biodiversity conservation for the presence of rare and endangered species that can be locally observed, such as *Paeonia mascula* (Fig. 2–44), listed in the Red Regional List (Conti et al., 1992). Also of great relevance to biodiversity conservation are the Palestine oak open forests, where it is often possible to observe important herbaceous species, such as *Stipa austroitalica* (Fig. 2–45), priority species of the Annex II of Habitats Directive (Domina & Santangelo, 2013), and numerous orchid species (Fig. 2–46).

Fig. 2–44. Paeonia mascula *details observed in mixed wood of* Quercus virgiliana *and* Quercus calliprinos *(Grottagiglio Wood – Cassano delle Murge) (Photo Rocco Carella).*

Fig. 2–45. Stipa austroitalica *observed in pastures in contact with Palestine oak maquis in the Bari hinterland (Photo Rocco Carella).*

Fig. 2–46. Ophrys fuciflora ssp. apulica *in clearings of mixed forest of* Quercus trojana *and* Quercus virgiliana, *with* Quercus calliprinos *in the dominated layer. Mesola Forest. Photo Rocco Carella).*

Finally, the Palestine oak in the study area shows a great ability to colonize poor and rocky soils, enhancing significantly one of the regional districts with the scarcest presence of spontaneous vegetation due to its morphology and historical agricultural vocation (Musca, 1987). The residual maquis and forest patches scattered in a territory as deeply altered by the anthropogenic impact as the Bari basin, could also provide fundamental resources to face the threats related to climate change in a territory dominated by crops. The recent, actual, quick decline of olive trees related to *Xylella fastidiosa* can also be seen as further evidence of climate change and, in this dramatic scenario, more varied fields with, the presence of relic patches of spontaneous vegetation, are predicted to be more resilient than monocoltures (Guinness & Whalpole, 2012).

References

ARMENISE V. 1952. *Bosco diradato, detto di Trullo Grotta del Lupo presso Cassano (Bari).* Ed. Centro per lo studio della flora e della vegetazione (Firenze): 468–474.

AZZAROLI A. & VALDUGA A. 1967. *Note illustrative della carta geologica d'Italia Foglio 177 "Bari" e 178 "Altamura".* Servizio Geologica d'Italia. Roma.

BALDACCI O. 1962. *Le Regioni d'Italia, XIV Puglia.* Edizioni UTET.

BIANCO P., BRULLO S., MINISSALE P. & SPAMPINATO G. 1998. *Considerazioni fitosociologiche sui boschi a Quercus trojana Webb della Puglia (Italia meridionale).* Studio Geobotanica, 16: 33–38.

BIANCO P., SCHIRONE B. & VITA F. 1989. *Considerazioni sulla distribuzione della Quercia spinosa in Puglia.* Acc. Ital. Sc. Forest., XXXVIII: 233–261.

BIANCO P., CASTELLANO M.A., PIRO G. & SCHIRONE B., 1981–1982. *Sulla distribuzione della Quercia spinosa in Puglia.* Ann. Fac. Agr. Univ. Bari, 32: 255–291.

BIONDI E., CASAVECCHIA S., GUERRA V., MEDAGLI P. BECCARISI L & ZUCCARELLO V. 2004. *A contribution toward the knowledge of semideciduous and evergreen woods of Apulia (south-eastern Italy).* Fitosociologia, 41(1): 3–28.

BRAUN-BLANQUET J. 1951. *Les groupements végétaux de la France Méditerranéenne.* Montpellier.

CAMUS A., 1936–1938. *Les Chénes. Monografie du genre Quercus.* P. Lechevalier Ed. Paris. Tome I: 434–463.

CARELLA R. 2017. *L'ultimo paradiso. Natura e paesaggio del Torrente Picone e dei suoi due bracci Torrente Baronale e Torrente La Badessa*. Aracne Edizioni.

CARELLA R. 2015. *The role of suburban area for biodiversity conservation: the case study of one of the biggest southern Italian cities*. Journal of Landscape Management, 6(2): 7–16.

CARELLA R. 2014. *I boschi dell'area di Mesola*. Dendronatura, 34(1): 96–104.

CARELLA R. 2013A. *Boschi e biodiversità nella Bassa Murgia Barese. Nuclei forestali d'interesse per la conservazione nel tratto alto del Torrente La Badessa.* Sherwood, 196: 46–50.

CARELLA R. 2013B. *Lama La Fitta, un solco erosivo dell'hinterland di Bari.* Umanesimo della Pietra, Riflessioni, 36: 127–134.

CARELLA R. 2013C. *Il tratto alto di Lama Cutizza nell'hinterland barese.* Dendronatura, 34(1): 96–104.

CARELLA R. 2012. *Lama Giotta, spartiacque tra il Nord-Ovest e il Sud-Est barese.* Umanesimo della Pietra. Riflessioni, 35: 153–158.

CARELLA R. 2011. *Ritrovamento di un boschetto di Quercus trojana Webb nel settore nord-occidentale delle Murge, in territorio di Cassano delle Murge. Considerazioni sulla distribuzione naturale del fragno sull'altipiano Murgiano.* Dendronatura, 32(2): 74–80.

CARELLA R., VITA F. & TRAVERSA A. 2003. *Lembi residui di vegetazione spontanea lungo Lama San Giorgio in Provincia di Bari*. Umanesimo della Pietra. Riflessioni, 26: 157–162.

CONTI F., MANZI A. & PEDROTTI F. 1992. *Libro Rosso delle Piante d'Italia.* Ministero dell'Ambiente. WWF Italia, Società Botanica Italiana.

CHIESURA LORENZONI F., CURTI L., LORENZONI G.G., LUCATO A. & MARCHIORI S. 1974. *Ricerche fitosociologiche sulle cenosi a Quercia spinosa del Salento (Puglia).* Not. Fitosoc., 8; 45–64.

CHIESURA LORENZONI F. & LORENZONI G.G. 1979. *Importanza delle cenosi a quercia spuinosa nella protezione del suolo e nella ricostituzione della vegetazione in Puglia.* Atti VIII Simp. Naz. Cons. nat.. Bari, Aprile: 413–422.

CRIVELLARI D. 1950. *Inchiesta sulla distribuzione del genere* Quercus *in Puglia.* Nuovo Giornale Botanico Italiano, n.s., vol. LVII, n.3: 335–350.

DOMINA G. & SANTANGELO A. 2013. *Stipa austroitalica.* The IUCN Red List on Threatened Species. Accessed 12 June 2016.

FORTE L., PERRINO E.V. & TERZI M. 2005. *Le praterie a Stipa austroitalica Martinovsky ssp. austroitalica dell'Alta Murgia (Puglia) e della Murgia Materana (Basilicata).* Fitosociologia, 42(2): 83–103.

GAVIOLI O. 1935. *Sulla dispersione del genere Quercus in Lucania.* Forlì: 105–124.

GENTILE S. 1962. *Frammenti di macchia con Quercus calliprinos Webb nel territorio di Ragusa (Sicilia Meridionale-Orientale).* Giorn. Bot. Ital., 110: 89–115.

GENTILE S. & GASTALDO P. 1976. *Quercus calliprinos Webb e Quercus coccifera L.: Ricerche sull'anatomia fogliare e valutazioni tassonomiche e corologiche.* Giorn. Bot. Ital., 110: 89–115.

GHAZAL A. 2008. *Landscape Ecological, Phytosociological and Geobotanical Study of Eu-Mediterranean in Syria.* Ph.D. Thesis, Faculty of Agricultural Science, University of Hohenheim, pp. 260.

GHAZANFAR S.A. & FISCHER M. 1998. *Vegetation of the Arabian Peninsula.* Springer Science+Business Media Dordrecht: 129.

GIANGUZZI L. & LA MANTIA (a cura di) AZIENDA FORESTE DEMANIALI – REGIONE SICILIA, 2000. *La Riserva di Monte Cofano.* Gruppo VIII Direzione Azienda FSDD.R.S.: 151.

GIANNUOLI S. 1951. *Un piccolo bosco a Quercus coccifera presso Mottola (Taranto).* Nuovo Giorn. Bot. Ital. n.s.: 603–606.

GUINNESS P. AND WALPOLE B. 2012. *Environmental Systems and Societies for the IB Diploma.* Cambridge University Press, p. 184.

LA MANTIA A. & GIANGUZZI L. 2003. *Considerations on the protection and forestal restoring of the Quercus calliprinos vegetation in Sicily.* Bocconea 16(2): 823–829.

LORENZONI G. & CHIESURA LORENZONI F. 1987. *Spiny oak (Quercus coccifera s.l.) coenoses of the Otranto coast (Lecce . South Apulia, Italy).* Acta Bot. Croat., 46: 85–64.

MACDONALD M.A 2003. *The role of corridors in biodiversity conservation in production forest landscape: a literature review.* Tasforest, 14: 41–52.

MONTERO DE BURGOS J.L. & GONZALES REBOLLAR J.L. 1973. *Diagramas bioclimatiquos.* Instituo Nacional para la Conservación de la Naturaleza, Madrid.

MUSCA G. 1987. *Storia della Puglia. Vol. 2.* Adda Editore, p. 34.

NAHAL I. 1961. *La garrigue a Quercus calliprinos Webb et Pistacia palaestina Boiss (Pistacio-Quercetum calliprini) de la montagne des Alouites de Syrie (R.A.U.).* Annales de l'école nationale des eaux et forêts, 18(3): 409–427.

PAFFETTI D., VETTORI C. & GIANNINI R. 2001. *Relict populations of* Quercus calliprinos *Webb on Sardinia island identified by chloroplast DNA sequences.* Forest Genetics, 8(1): 1–11.

PIGNATTI S. 2002. *Flora d'Italia,* Vol. 1. Edagricole, Bologna.

RIVAS MARTINEZ S. 2004. G*lobal bioclimatics. (Clasificación bioclimática de la Tierra*): Version 27–08–04. Phytosociological Research Center. Madrid. University Comutense of Madrid.

ROMANAZZI A. 2015. *Modelling and management of a Mediterranean karstic coastal aquifer under the effects of seawater intrusion and climate change.* Environmental Earth Science, 74(1):115–128.

SABATO S. 1972. *Considerazioni sul significato fitogeografico ed ecologico di* Quercus coccifera *L. s.l. nel Salento (Puglia).* Webbia, 27(2): 517–549.

SERRUYA C. 1978. *Lake Kinneret.* Monographiae Biologicae, 32: 7–13.

SISTEMA DE CLASIFICACIÓN BIOCLIMÁTICA MUNDIAL, 1996–2009, S. Rivas-Martinez, Y.S. Rivas Sáenz, Centro de Investigaciones Fitosociológicas, España. http://www.ucm.es/info/cif. Accessed 18 February 2016.

THORNTHWAITE C.W. & MATHER J.R. 1955. *The water balance.* Publications in Climatology, New Jersey, Drexel Institute of Technology.

TOUMI L. & LOUMARET R. 2010. *Genetic variation and evolutionary history of holly oak: a circum-Mediterranean species complex (Quercus coccifera L./Quercus calliprinos (Webb) Holmboe, Fagaceae).* Plant Syst. Evol., 290: 159–171.

TUTIN T.G., HEYWOOD W.H., BURGES N.A., VALEBTINE D.H., WALTER

S.M. & WEBB D.A. (EDS) 2010. *Flora Europaea.* Vol. 1, 2nd edition. Cambridge University Press, Cambridge.

VITA F., CARELLA R., D'EMERICO S. 2001. *Analisi bioclimatica e fitosociologica delle macchie-foreste a dominanza di Quercus coccifera L. dei dintorni di Bari.* Monti e Boschi, 52(1): 50–58.

VITA F., PROCINO V., CARELLA R. & D'EMERICO S. 2002. *Analisi bioclimatica e vegetazionale di un bosco a dominanza di Quercus pubescens Willd s.l.*. Monti e Boschi, 52(3–4): 19–25.

ZOHARI M. 1961. *On the oak species of the Middle East.* Bull. Res. Council Israel, Sect. D., Bot. 9(4): 161–186.

ZOHARI M. & ORSHAN G. 1965. *An outline of the geobotany of Crete.* Suppl. Israel J. Bot: 1–49.

CHAPTER 3

THE IMPORTANCE OF *QUERCUS CALLIPRINOS* TO BIODIVERSITY AND LANDSCAPE CONSERVATION IN THE MEDITERRANEAN BASIN

ROUBINA GHATTAS

3.1 Overview: the Mediterranean forests

The Mediterranean environment supports various ecosystems depending on rock-soil type, climatic conditions, and land-use history. In general, plant communities of this region are composed of two dominant components: annual and perennial herbaceous plants and sclerophyllous, mostly evergreen, woody species including dwarf shrubs, shrubs, and trees. Woodlands that are mostly dominated by the *Quercus* species are the common vegetation formations in the Mediterranean region. The Mediterranean vegetation cover is known for its rapid recovery in case of natural disaster or human interference. The vegetation succession, specifically the development from herbaceous species into low shrubs and later to woody species, is a regular process that takes place in the Mediterranean region (Perevolotsky & Seligman, 1998).

The forested areas in the Mediterranean region are well known for their species wealth, as they are supported by major bio-geographic diversity, which is a result of the diverse physical and human contexts that have shaped the Mediterranean region for nearly 10,000 years (Thirgood, 1981; Pons & Quézel, 1985). It is a treasure chest of biodiversity, which hosts a large variety of plants and has been identified with high genetic diversity for a wide range of crops, such as wheat, lentils, peas and vetch, which were domesticated 10,000 years ago. Plants of this region are of greater importance to mankind than those of most parts of the world. The life-sustaining crops of wheat, barley, vines, olives, onions and pulses and valued holy trees all originated within the geographical land of the Mediterranean region. This is partly due to its much dissected ecosystems, with its consequent extremely wide range of agro-ecological conditions.

3.2 *Quercus calliprinos* as a typical Mediterranean sclerophyllous species

Quercus L. (Oak) is one of the exceptionally important woody species worldwide. *Quercus calliprinos,* the Palestine oak, in particular, is a typically Mediterranean evergreen tree species of the maquis of the region, and the major evergreen element in the *Q. calliprinos – Pistacia palaestina* association of the Mediterranean region. It is an oak classified as part of the *Cerris* section of the species. It is native to the eastern Mediterranean region and south-west Asia, and grows throughout the Middle East (Zohary, 1960).

The Mediterranean Woodland and Forest eco-region in northern Africa supports more the growth of specific types of evergreen oaks, including cork oak (*Quercus suber*) forests; holm oak (*Quercus ilex*) and holly oak (*Quercus coccifera*) forests; and wild olive (*Olea europaea* and *O. maroccana*) and carob (*Ceratonia siliqua*) woodlands and maquis. The Mediterranean woodland and forest eco-region in southern Europe, on the other hand, supports the growth of other types, such as Portuguese oak (*Quercus faginea*), Italian oak (*Quercus frainetto*), Pyrenean oak (*Quercus pyrenaica*), Turkey oak (*Quercus cerris*), Aleppo oak (*Quercus infectoria*), cork oak (*Quercus suber*) and others.

Quercus calliprinos is found on the Terra Rossa soil, with dominant parent materials of dolomite and hard limestone. They usually appear as a large scrub, although some tall individual trees do exist, particularly those that were treasured by local residents as "holy trees" and hence remained untouched for many years. The trees are relatively small and have only one trunk. On the same tree are male and female flowers, where wind pollination takes place (Modzelevich, accessed 7 July 2023).

Fig. 3–1. Pistacia palaestina *and* Quercus calliprinos *typical coenoses (Al Qarin protected area) (Photo courtesy the Applied Research Institute-Jerusalem, 2016).*

Q. calliprinos is known to cover vegetation zones up to 350 metres above sea level and in some areas, such as Lebanon, up to 1,000 metres (Karam et al., 2011). The co-dominant species associated with *Q. calliprinos* are *Pistacia palaestina* (Fig. 3–1), *Crataegus azarolus*, *Arbutus andrachne*, *Phillyrea media*, *Ceratonia siliqua*, *Pyrus syriaca*, *Styrax officinalis*, *Rhamnus palaestinum* (Ghattas et al., 2013), and *Juniperus oxcedrus* (Karam et al., 2011). It is sometimes accompanied by other shrub and herb species, such as *Asparagus aphyllus*, *Smilax aspera*, *Clematis cirrhosa*, *Lonicera etrusca*, *Cistus salviifolius*, *Cistus incanus*, *Calicotome villosa*, *Sarcopoterium spinosum*, *Thymus capitatus*, *cyclamen persicum*, *Salvia fruticosa*, *Teucrium polium*, *Teucrium divaricatum*, *Salvia dominica*, and other plants (Ghattas et al., 2013). At other times, as observed in some Palestinian forest areas (Wadi Al Quff protected area), *Quercus calliprinos* occurs in mixed broadleaf/coniferous forest, where the pine trees form the upper storey of the forest, while the species with other evergreen shrubs form the lower.

3.3 Role of *Quercus calliprinos* in maintaining biodiversity and landscapes

Oak forests are an important ecosystem type of the natural vegetation in the Mediterranean region, providing a wide range of ecosystem functions and services (Ya'ar Bar'am, 2009). In the disturbed landscapes occurring in the Mediterranean region, the well-kept areas, especially those with holy woods, form natural bases with significant biodiversity elements. Sacred trees, groves and forests, such as the evergreen oak forest, are found all over the Mediterranean basin. They were established, sometimes centuries ago, for spiritual and cultural purposes (Harding et al., 1998). This holy protected evergreen oak tree, dominantly *Q. calliprinos*, supports the conservation of a great number of associated species (Martinez & Amar, 2014), even small species such as ants, spiders and snails, especially in hot regions, by providing suitable habitats and shelter. This is manifested in Palestine (in Um Er Rihan, A'tara, Bili'n, Ein Qiniya and others), Lebanon (Mazarait Beit Slaibi, Rait, and others) (Karam et al., 2011), Syria (in Baer, Djebel Ansarieh and others) (Rolley, 1948) and other Mediterranean areas. In conclusion, holy evergreen oak forests have a high conservation value, because they are rare examples of intact Mediterranean forests, representing important elements in ecosystems with stable species composition and high cultural values (Deil et al., 2009).

Evergreen oaks are drought-resistant species, even in areas facing induced degradation, such as many sites in the Mediterranean ecosystem. Evergreen oaks are known to maintain a high transpiration rate during periods of intense drought; hence oaks quickly respond to drought stress at different seasons; it acts as a water saver and benefits from efficient water loss control capabilities (Sperlich et al., 2015).

As was noted in a study on the "eco-physiological behaviors of two woody oak species to combat desertification in the east Mediterranean" by Dr F. Karam (2011), *Q. calliprinos* is a species that keeps water and nutrients in dry conditions, especially because of reduced plant competition in degraded sites. It was also noted that the rate of seasonal variations in water use efficiency (WUE) values shows tolerance to drought plant species where the WUE decreases up to 40% in *Q. calliprinos* in healthy sites and up to 33% in degraded site; thus *Q. calliprinos* can be used as an indicator for degraded areas induced by climate or anthropogenic factors. It will be the last *Quercus* species to disappear when rainfall is lacking.

Hence *Q. calliprinos* dominates the Mediterranean maquis forest where drought conditions are prevalent, which makes the evergreen oaks a moderator for the effect of drought by keeping water in its root and providing shade and a micro-habitat for other plant and animal species. The falling leaves become compost that supports the growth of microorganisms and hence increases the fertility of the soil. Evergreen oaks support a greater biodiversity than any other tree in the Mediterranean; as oak forests provide a habitat rich in biodiversity, they support more life forms than any other native trees. Its ecological importance is as a habitat, shelter and food source for much wildlife in the Mediterranean forest. It hosts hundreds of species of invertebrates, including reptiles, insects, molluscs and arachnids, supplying many birds with an important food source. In autumn, mammals such as badgers, deer, foxes, rodents and wild boars take advantage of the falling acorns. The old trees can be a habitat to many insects, such as beetle fauna, which grow and live in trees hollows (Erol et al., 2012). Some studies indicate that old oaks are an important habitat to rich fauna, with many species new to science (Schillhammer et al., 2007; Sama et al., 2011).

Correspondingly, Mediterranean maquis trees show high resilience to extreme disturbance, as they survive canopy removal. Some research results indicate that the absolute regrowth rate of the dominant evergreen *Q. calliprinos* after cutting and under grazing is higher than its accompanying deciduous species (Hare'l & Gidi, 2011). Thus, this tree is at a clear advantage for the Mediterranean basin, especially as it has the ability to acclimatize to the long human disturbances known to take place in the region (Naveh, 1990). Apparently, large oak trees possess higher carbohydrate reserves in their roots, which support their vigorous post-disturbance regrowth (Canadell & Lopez-Soria, 1998), in addition to the large number of stems, characteristic of *Q. calliprinos* tree, for survival and regrowth after a disturbance.

Another fact is that for *Q. calliprinos,* in particular, its boughs are often readily grazed by livestock and can constitute a significant part of their diet. Thus, natural oaks can serve as pasture for cattle and goats (Perevolotsky et al., 1992). Since the production of herbaceous species in Mediterranean woodland is relatively low, oak leaves and young stems are an important food source for animals that graze this ecosystem. Oak leaves are important, especially in summer and fall when the dry herbaceous plants have very limited nutritional value[1] (Brosh et al., 2006). It is worth noting that oaks leaves provide energy but not the daily protein requirements for herds (Perevolotsky et al., 1993). Grazing the oaks during summer gives a relative advantage to some deciduous tree species (Hare'l & Gidi, 2011).

Recently, it has been acknowledged that biodiversity will be best conserved if the closed woody landscape possesses open patches (Perevolotsky & Seligman, 1998). Grazing by livestock is one of the efficient tools for keeping the woodlands heterogeneous in structure, but it should not be forgotten that species richness will decrease under either heavy grazing pressure, early grazing and/or a long grazing period (Sterenberg et al., 2000). Thus, to improve habitat heterogeneity and increase species richness, managed grazing regimes, with regular mowing of that habitat, is recommended (Dahms et al., 2010). Such regimes are found to be adequate to be implemented in the Mediterranean maquis dominated by *Quercus calliprinos.*

Q. calliprinos provides special and sustainable habitats, since it is palatable to a certain extent and resistant to grazing, disturbances and drought conditions, especially when accompanying a number of plant and animal species that are more susceptible to natural and human disturbances. This also secures a balanced Mediterranean maquis with a diverse number of species.

---

[1] The goat is the domestic animal most commonly found in areas in which a woody element is dominant.

3.4 Beneficial uses of *Quercus calliprinos*

Q. calliprinos is highly respected across most cultures in the Mediterranean basin. It has for centuries been a symbol of strength and survival. The evergreen oak was holy to many gods, including Zeus (Greek), Jupiter (Roman) and Dagda (Celtic). Each of these gods ruled over thunder and lightning; oak trees are prone to lightning strikes as they are often the tallest living feature in the landscape (Foster, 2016). Worship often took place in oak groves. Royalty has had a long association with oak trees too; ancient kings decorated themselves with crowns of oak leaves, especially in victory parades. It is mentioned often in the Bible as a symbol for strength, a metaphor for power and might, and as a place of worship.

Q. calliprinos produces one of the hardest and most durable timbers among Mediterranean trees. However, it is ready to be used in construction after it is only 150 years old. Its wood has been used right up to today in art for sculpting statues and ornamentation, for furniture, construction (flooring), industry (shank for the plough, ox, cane for the elderly) and for the production of coal. It is widely used, especially in Europe, for the production of wine barrels, and its cork for the production of bottle stoppers (Woodland Trust, n.d.). Oaks are also known as important ornamental trees.

Traditionally, the leaves, bark and acorns were believed to heal many medical conditions, including diarrhea, colds, fever, coughs and bronchitis, to tackle inflammation and kidney stones, and for stimulating the appetite and improving the digestion (WebDM). Powerful antibacterial, antifungal, antioxidant, parasite-fighting or immune-boosting substances have been found in native plants such as Palestine oak (Nyerges, 1997). Historically, humans also collected acorns (Fig. 3–3) and processed them into flour for bread making. These culinary techniques have mostly died out following the domestication of wheat production 10,000 years ago, leaving the acorn harvest for wild birds and mammals. Tannin found in the bark has been used to tan leather since at least Roman times.

Fig. 3–2. Quercus calliprinos. *Jenin Governorate (Photo courtesy the Applied Research Institute-Jerusalem, 2016).*

Fig. 3–3. Quercus calliprinos *acorn (Photo courtesy the Applied Research Institute-Jerusalem, 2016).*

3.5 The conservation of Mediterranean forests and the impact on *Quercus calliprinos*

The conservation of forests and forest species in the Mediterranean basin is a complex issue, given the wide range of ecological conditions, diverse uses and the pressure exercised by different cultural groups. The management of forest systems demands consideration not only of forest communities and species, but also of associated ones, especially on the edges of forests and in clearings. Although the fairly widespread distribution of most forest species would indicate that any threat to them is limited, the fair number of trees representing key species, such as *Quercus*, could be vital for the sustainability of ecosystem functions.

Fig. 3–4. Quercus calliprinos *young plant (Photo courtesy the Applied Research Institute-Jerusalem, 2016).*

Many woody species can subsist in isolated populations and in very restricted areas for several decades, as long as the main biotic conditions are sustained. However, geographically isolated and reduced populations are more vulnerable than any others and suffer more from the impact of random environmental events. Frankel, Brown & Burdon (1995) place the minimum number of trees necessary for long-term maintenance of a viable forest population at about 500. Therefore, it seems best to set up systems incorporating large unbroken reserves (of at least several hundred hectares) surrounded by buffer zones, making up a mosaic with differently managed areas (Frankel et al., 1995). Such structures are foreseen in the "Biosphere Reserves" of UNESCO's Man and the Biosphere Programme.

As discussed above, extensive deforestation has occurred throughout the Mediterranean region. Human impact in the region is very high, because it hosts most of the north African and eastern Mediterranean human urban areas, and those countries are facing strong socio-economic and political instability. High population growth results in the rapid change in land uses and intense land conversion, mainly for agriculture, urban development, industry and quarries. Other disturbances are traditional agricultural activities, such as cutting and grazing. Grazing may help in maintaining the genetic diversity within plant populations. But under heavy grazing pressure, grazing may delay the vegetation succession process. Within dense stands of woodland dominated by *Quercus calliprinos*, the herbaceous species diversity is negatively affected. Therefore, an appropriate management regime must be implemented. In relation to protecting landscape and species diversity in the Mediterranean ecosystems, when cutting is applied in combination with grazing, special care must be paid to trees other than *Q. calliprinos* that are more negatively affected by such disturbances (Sternberg et al., 2000); summer grazing might give a relative advantage to some deciduous tree species (Hare'l & Gidi, 2011). Finally, the habitats and populations of certain species, such as oak species, should be adequately represented and conserved in protected areas to ensure their long-term survival and thus the maintenance of the ecological processes related to them.

References

BERRAHMOUNI N. & REGATO P. n.d. *Northern Africa: Morocco, Algeria, and Tunisia.* WWF. http://www.worldwildlife.org/ecoregions/pa0513.

BROSH A., HENKIN Z., ORLOV A. & AGARONI Y. 2006. *Diet composition and energy balance of cows grazing on Mediterranean woodland.* Livestock Science issue, 102: 11–22.

CANADELL J. & LOPEZ-SORIA L. 1998. *Lignotuber reserves support regrowth following clipping of two Mediterranean shrubs*. Funct. Ecol., 12: 31–38.

DAHMS H., LENOIR L., LINDBORG R., WOLTERS V. & DAUBER J. 2010. *Restoration of semi-natural grasslands: What is the impact on ants?* Restor Ecol, 18: 330–337.

DEIL U., FROSCH B., JÄCKLE H., MHAMDI A. & ACHHAL A. 2009. *A geobotanical analysis of forest patches on sacred sites in Northern Morocco.* Berichte der Reinhold-Tüxen-Gesellschaft, 21: 174–195.

EROL A., NICKOLAS J. & TOLGA G. 2012. *Saproxylic beetles on old hollow oaks (Quercus spp.) in a small isolated areas in southern Turkey.* Zoology in the Middle East issue, 57: 105–114.

FOSTER T., 2016. *The Tree of Life. Mature Times: the voice of our generation.* http://www.maturetimes.co.uk/tree-life/.

FRANKEL O.H., BROWN A.H.D. & BURDON J.J. 1995. *The conservation of plant biodiversity*. Cambridge University Press. Cambridge. pp. 299.

GHATTAS R., BANNOURA E. & KHOURI, S. 2013. *Sustainable Use of Palestinian Forest Resources.* The Applied Research Institute-Jerusalem (ARIJ). Palestine. pp. 52.

HARDING J.S., BENFIELD E.F., BOLSTAD P.V., HELFMAN G.S. & JONES EBD III. 1998. *Stream biodiversity: the ghost of land use past.* Proc Natl Acad Sci, 95: 14843–14847.

HARE'L A. & GIDI N. 2011. *Quercus calliprinos regrowth advantage under grazing in Mediterranean maquis and its management implications.* Forest Ecology and Management, 261: 143–147.

KARAM F., DOULIS A., OZTURK M., DOGAN Y. & SAKCALI S. 2011. *Eco-physiological behaviors of two woody oak species to combat desertification in the east Mediterranean – a case study from Lebanon.* The 2nd International Geography Symposium GEOMED 2010. Elsevier, Procedia Social and Behavioral Sciences, 19: 787–796.

MARTINEZ J. & AMAR Z. 2014. *The preservation value of tiny sacred forest of the oak Quercus calliprinos and the impact of livestock presence.* Insect conservation, 18(4): 657–665.

MODZELEVICH, M. *Flowers in Israel E-database.* http://www.flowersinisrael.com. Visited on 7th July 2023.

NAVEH Z. 1994. *The role of fire and its management in the conservation of Mediterranean ecosystems and landscapes.* In J.M. Moreno & W.C. Oechel (eds), *The Role of Fire in Mediterranean –type Ecosystems.* Springer-Verlag. New York: 163–185.

NAVEH Z. 1990. *Ancient man's impact on the Mediterranean landscape in Israel: ecological and evolutionary perspectives.* In S. Bottema, G.Entijes-Neiborg & W. Zeit (eds), *Man's Role in shaping East Mediterranean Landscape*. Balkema. Rotterdam: 43–52.

NYERGES C. 1997. *Natural Remedies: The Health Benefits of Oak and Elder Trees.* Mother Earth News. http://www.motherearthnews.com/natural-health/health-benefits-of-oak-and-elder-trees-zmaz97jjzgoe.aspx.

PEREVOLOTSKY A., BROSH A., EHRLICH O. & HOLZER Z. 1993. *Nutritional value of common oak (Quercus calliprinos) browse as fodder for goats: Experimental results in ecological perspective.* Small Ruminant Research, 1(2): 95–106.

PEREVOLOTSKY A. & SELIGMAN N.G. 1998. *Degradation of Mediterranean rangeland ecosystems by grazing: Inversion of a paradigm.* Bioscience, 48: 1007–1017.

PONS A. & QUÉZEL P. 1985. *The history of the flora and vegetation and past and present human disturbance in the Mediterranean region.* In C. Gomez-Campo (ed.), *Plant conservation in the Mediterranean area.* Geobotany, 7. W. Junk. Dordrecht, the Netherlands: 25–43.

ROLLEY J. 1948. *Forest conditions in Syria and Lebanon.* FAO Document Repository. Unasylva, 2: 77–80.

SAMA G., JANSSON N., AVCI M., SARIKAYA O., COSKUM M., KAYIS T. & OZDIKMEN H. 2011. *Preliminary report on a survey of the saproxylic beetle fauna living on old hollow oaks (Quercus spp.) and oak wood in Turkey.* Munis Entomology and Zoology, 6: 819–831.

SCHILLHAMMER H., SNALL S., COSKUN M. & JANSOON N. 2015. *The West Palearctic species of Hesperus Fauvel, 1874, with description of three new species from Turkey.* Koleopterologische Rundschau, 77: 123–132.

SPERLICH D., CHANG C.T., PENELAS J., GRACIA C. & SEBASTE S. 2007. *Seasonal variability of foliar photosynthetic and morphological traits and drought impacts in a Mediterranean mixed forest.* Oxford journals, tree physiology, 35(5): 501–520.

STERENBERG M., GUTMAN M., PEREVOLOTSKY A., UNGAR E.D. & KIGEL L., 2000. *Vegetation response to grazing management in a Mediterranean herbaceous community; A functional group approach.* Journal of Applied Ecology, 37: 224–237.

THIRGOOD J.V. 1981. *Man and the Mediterranean forest.* Academic Press. New York, pp. 194.WEBMD. OAK BARK OVERVIEW INFORMATION. http://www.webmd.com/vitamins-supplements/ingredientmono-498-oak%20bark.aspx?activeingredientid=498&activeingredientname=oak%20bark.

WESTPHAL C., VON OHEIMB G., MEYER-GRUNEFELD M., TREMER T., HARDTLE W., LEVANOMY T., DAYAN T. & ASSMAN T. 2009. *An old*

Quercus calliprinos forest of high nature conservation value in the Mediterranean region of Israel. Conservation Studies in Israel, 57(1–2) 13–23.

WOODLAND TRUST. OAKS: ENGLISH (QUERCUS ROBUR). https://www.woodlandtrust.org.uk/visiting-woods/trees-woods-and-wildlife/british-trees/native-trees/english-oak/.

WORLD WILDLIFE FUND. 2001. *The Mediterranean Forests. A new conservation strategy.* WWF. MedPO, Rome.

ZOHARY M. 1990. *The maquis of Quercus calliprinos in Israel and Jordan.* Bulletin of the Research Council of Israel, 9(2): 51–72.

ZOHARY M. 1973. *Geobotanical foundations of the Middle East*. Gustav Fischer Verlag. Stuttgart. 2 vol. (X, XI), 738 p.

ZOHARI M. 1966a. *On the oak species of the Middle East*. Bull. Res. Counc. Isr. Sect. D, 9: 167–186.

ZOHARI M. 1966b. *Flora Palaestina*. The Israel Academy of Sciences and Humanities, Jerusalem, 1: 18–329.

ZOHARI M. 1962. *Plant life of Palestine, Israel and Jordan*. Ronald Press. New York: 262 pp.

ZOHARI M. 1961. *On the oak species of the Middle East*. Bull. Res. Counc. Isr., 9(4): 161–186.

ZOHARI M. 1960. *The maquis of Quercus calliprinos in Israel and Jordan*. Bull. Res. Counc. Isr., 9D: 51–72.

CHAPTER 4

ECOLOGY OF KERMES OAK AND POTENTIAL THREATS FACED BY IT RELATED TO CLIMATE CHANGE

MUNIR OZTURK AND VOLKAN ALTAY

4.1 Introduction

The genus *Quercus* L. is an ecologically diverse and economically important plant species (Simeone et al., 2016; Ozturk & Altay, 2021). It has been valued throughout its existence, especially for its hardwood. There are nearly 600 species of this genus distributed across the northern hemisphere (Ozturk & Altay, 2021).

Quercus, commonly known as "Oak", belongs to the family *Fagaceae*. It includes many woody plants, mostly economically important shrubs or trees, able to survive for centuries, possessing simple alternate leaves; the flowers are wind-pollinated and the fruit is the acorn. The plants of this genus are one of the most imperative clades of woody angiosperms with regard to species diversity and ecological dominance. The oaks are dominant members of a wide variety of habitats, including temperate deciduous, temperate, subtropical evergreen forests, as well as sub-tropical and tropical savannah and subtropical woodlands. The species of this genus are found in a wide variety of urbanized and semi-urbanized environments, preferring loamy, well-drained soils. The extensive roots of these plants connect to over three times the height of the tree, extending down to 4.6–12.2 metres. These plants fluctuate in size from small shrubs to lofty trees (Tantray et al., 2017).

Quercus coccifera L. (Fig. 4–1) is one of the most common evergreen species of the Mediterranean maquis flora (Tsiourlis et al., 2009; Babalik et al., 2021). In several taxonomic studies many synonyms of Kermes Oak (*Q. coccifera)* have been reported. An updated list, taking into account the 2023 online systematic database (www.powo.science.kew.org), is presented here.

Fig. 4–1. General view of Quercus coccifera. (Note: **A–C**: Young flowering branches and leaves; **D–F**: fruits and cupulas.)

4.2 Ecology of Quercus coccifera from Istanbul (Turkiye)

The native range of *Q. coccifera* lies in the Mediterranean region. It is a shrub or tree and grows primarily in the subtropical biome (www.powo.science.kew.org). This species is distributed throughout the Mediterranean region, from Palestine and to Portugal (Tsiourlis et al., 2009; Ozturk & Altay, 2021). It dominates the low and high Mediterranean shrub formations, mainly on limestone substrates (Tsiourlis et al., 2009; Ozturk & Altay, 2021). The distribution area includes Albania, Algeria, Bulgaria, Cyprus, France, Greece, Italy, Lebanon, Libya, Morocco, Palestine, Portugal, Syria, Spain, Tunisia and Turkiye (www.powo.science.kew.org).

Kermes oak is one of the important shrubs dominating the Mediterranean basin. The plant flourishes under typical Mediterranean climatic conditions, with long lasting summer droughts, a variety of soil types, acidic or basic parent materials, covering more than two million hectares (Babalik et al., 2021). In Turkiye, this plant is distributed throughout the Mediterranean coastal zone, extending quite far inwards at some places (Altay et al., 2012b; Ozyıgıt et al., 2015; Babalik et al., 2021).

In Istanbul (Turkiye), in the areas where maquis was destroyed by anthropogenic impacts, a total of 11 quadrate samples were recorded (Tab. 4-1). This plant community mainly belongs to the Mediterranean shrublands maquis and is represented in the localized parts of the study area. These present a degraded look due to heavy urbanization and a large number of representatives from the ruderal plant communities have invaded these areas due to the change in site conditions (especially species belonging to the classes *Stellarietea media* and *Artemisietea vulgaris*) (Altay, 2009; Altay et al., 2012a).

The kermes oak community is the next successional stage from *Arbuto unedonis-Phillyretum latifoliae*, with similar floristic composition, the only difference being that the oak prevails in the shrub layer. This community thrives on nutrient-rich, sandy loamy clay soils. The *Quercus coccifera* association evaluated here usually contain 4.66% organic matter, 10.62 kg/decare phosphorus, 1.60% $CaCO_3$, and 66.69 kg/decare potassium.

The soils of this association are generally sandy-clayey loam in texture, slightly acidic, nonsaline and less calcareous, containing high levels of organic matter, rich levels of phosphorus and potassium. It optimally develops from April until July, commonly occupying the slopes along the coastline (Altay, 2009; Altay et al., 2012a).

Q. coccifera dominant communities are common in different regions of Turkiye and are classified within different syntaxa. In the Black Sea region two communities from *Quercetalia ilicis* have been described: *Phillyreo-Quercetum coccifera* (Kutbay et al., 1999) and *Siderito dichotomae-Quercetum cocciferae* (Karaer et al., 1999). In the Aegean and Mediterranean region, the diversity of *Quercus coccifera* syntaxa is even greater (Ugurlu & Şenol, 2005).

The syntaxonomic structure of this *Quercus* association is summarized below:

- Quercetea ilicis Br.-Bl. ex A. Bolos et O. de Bolos in A. Bolos y Vayreda 1950
- Quercetalia ilicis Br.-Bl. ex Molinier 1934
- Arbuto andrachnes-Quercion cocciferae Barbero et Quezel 1979

Association: *Phillyreo latifoliae-Quercetum cocciferae* Özen 2010 (Tab. 4-1)

Character species: *Quercus coccifera, Asparagus acutifolius, Osyris alba, Paliurus spina-christi, Phillyrea latifolia, Pistacia terebinthus* ssp. *terebinthus, Ruscus aculeatus*, and *Teucrium chamaedrys* ssp. *chamaedrys.*

Quadrate number	1	2	3	4	5	6	7	8	9	10	11
Altitude (m)	16	9	6	65	72	52	247	20	15	234	7
Quadrate samples width (m^2)	35	35	40	100	100	100	50	100	60	80	35
Direction (^{0})	S	S	N	S	S	W	E	W	S	SW	N
Slope (%)	15	30	30	5	30	30	10	45	40	10	10
Plant cover (%)	85	85	95	95	90	100	90	95	98	95	95

Quercus coccifera	54	33	44	33	54	54	55	44	44	44	54
***Quercetea ilicis (-etalia)* Class**											
Phillyrea latifolia	11	11	12	22	12	11	11	11	22	22	22
Asparagus acutifolius	11	11	12	12	22	11	11	11	12	12	11
Osyris alba	11	11	12	12	12	12		11	11		
Ruscus aculeatus var. *angustifolius*	11	11		11				11	11		
Pistacia terebinthus subsp. *terebinthus*		22		11	11				11		
Rubia peregrina			+2				+2	+1	+2		
Calicotome villosa	12		11					12			
Crataegus monogyna				11		12				11	
Paliurus spina-christi				12	12				11		
Olea europaea				11					11		
Arbutus unedo										11	
***Quercetea pubescentis* Class**											
Quercus pubescens	12	12				12		11			
Rosa canina				11		11		11	11		
Teucrium chamaedrys	+2	+1				+2					

Brachypodium sylvaticum						+1		+1			
Laurus nobilis						11				11	
Tamus communis				+2							
Hypericum perforatum					11						
Quercus infectoria						11					
Hedera helix						+1					
***Cisto - Micromerietea* Class**											
Cistus creticus	11	11	12		12						12
Cistus salviifolius				12		12	12	12			
Sarcopoterium spinosum	12			11				11			
Spartium junceum		11			11			12			
Dorycnium pentaphyllum subsp. *herbaceum*			+2					+1			+2
Psoralea bituminosae			11	12							12
Melica ciliata subsp. *ciliata*				+2	+2						
Erica arborea						11		11			
Pteridium aquilinium						12					
Lavandula stoechas subsp. *stoechas*								11			

***Molinio - Arrhenatheretea* Class**											
Dactylis glomerata subsp. *hispanica*	+2	+2	+2			+2	+2	+1		+2	+2
Lolium perenne	+2	+1	+1	12							
Plantago lanceolata	11			12							
***Stellarietea mediae* Class**											
Pallenis spinosa	11	12	12	12	12						
Rapistrum rugosum	+1	+1		12					11		
Bromus hordeaceus	+2	+1									
Capsella bursa - pastoris				12	12						
Sonchus asper subsp. *glaucescens*							11			11	
Cynodon dactylon							+2			+2	
Plantago lagopus	+2										
Hordeum murinum subsp. *leporinum*									+2		
Portulaca oleracea										+2	
Senecio vulgaris										+1	

***Artemisetea vulgaris* Class**											
Foeniculum vulgare	+1			12							
Carduus nutans	12										
Scolymus hispanicus	11										
Centaurea solstitialis subsp. *solstitialis*				11							
Centaurea diffusa					11						
Picnoman acarna					12						
Galium aparine										+2	
Convolvulus arvensis										+2	
Cichorium intybus										11	
Companions											
Anthemis cretica subsp. *tenuiloba*	+1	+2	+2	12		+2	+2				
Stachys byzantinum	12	+2	+1	12	12						
Hordeum bulbosum	+2	11	+2	12						+2	
Agrostis capillaris				12	12		+1		12		
Rubus canescens				12		11			12		+2
Dianthus leptopetalus				12	12		+1				
Carlina corymbosa					12	11	11				

Scanbiosa columbaria subsp. *columbaria*	11			22							
Convolvulus cantabrica	+2	+2									
Muscari comosum	+2	+2									
Ornithogalum umbellatum	+1		+1								
Aegilops geniculata		+2	+1								
Nigella damascena				12	12						
Bellis perennis							+1	+1			
Cruciata taurica				+2							+2
Eryngium campestre var. virens				11						11	
Trifolium stellatum	+2										
Linum trigynum	+2										
Ferulago confusa		12									
Trifolium angustifolium			+1								
Briza media			+2								
Onobrychis oxyodonta			12								
Alcea pallida				12							
Tordylium apulum					12						
Scutellaria albida					12						
Daucus guttatus					11						

Cirsium polycephalum	11		
Berula erecta		+1	
Allium paniculatum subsp. *angustifolium*		+2	
Ballota nigra subsp. *anatolica*			22
Crepis sancta			+2

Tab. 4-1. Phillyreo latifoliae-Quercetum cocciferae *association.*

4.3 Potential threats faced by *Quercus coccifera*

The areas occupied by the oak species are facing overgrazing and overexploitation as well as wildfires, which change the physiognomy together with the composition of natural vegetation in the Mediterranean, thus appearing as the major factors responsible for the regional predominance of this species (Tsiourlis et al., 2009; Ozturk & Altay, 2021). The sclerophyllous and shrubby habit as well as multistemmed look attract less human attention. As a result, its resprouting ability enables coppicing and overgrazing. Its rapid post-fire regeneration lies behind the recovery success of *Q. coccifera* (Tsiourlis et al., 2009; Ozturk & Altay, 2021). This species follows a slow-growth strategy in general, which enables replacing the initially dominant fast-growing species through succession (Paraskevopoulos et al., 1994). In a study on grazing effects on different maquis trees (Tavşanoğlu & Coşkun, 2009), *Q. coccifera* has been found to have the least grazing index values, the reason being its vertical growth strategy enables it to avoid grazers (Perevolotsky & Haimov, 1992).

4.3.1 Land use changes

In most of the Mediterranean region, evergreen and deciduous broad-leaved forests represent the late-successional plant communities. The mosaic-like landscapes characterized by high structural and floristic diversity, in particular in the herbaceous and shrubby vegetation, have been shaped by the millennia of traditional land use (Naveh, 1998; Babalik et al., 2021; Ozturk & Altay, 2021). The almost complete destruction of the natural arboreal vegetation has been reported (Westphal et al., 2009; Ozturk & Altay, 2021). Often corresponding to the different stages of regressive succession, the original so-called natural forests left behind have fully changed (Simeone et al., 2016; Babalik et al., 2021; Ozturk & Altay, 2021).

We come across dramatic land use changes in the Mediterranean region with the abandonment of traditional agrosilvo pastoral practices in the north (Babalik et al., 2021). In the south we come across intensification of land use with heavy urbanization and modern agricultural practices. The shrublands develop in dense sclerophyllous oak forests following changes in the traditional land use, as examplied in the north-western Mediterranean (Babalik et al., 2021; Ozturk & Altay, 2021).

4.3.2 Drought

The species of the genus *Quercus* are facing a great threat from anthropogenic (deforestation) as well as environmental (biotic and abiotic stresses) pressures, endangering the sustainability and survival of the oaks species (Haavik et al., 2015; Natalini et al., 2016; Guerrero-Sanchez et al., 2022). According to Muhammad et al. (2023) oak species can compete in drought and high-temperature environments to some extent, but climate change with a higher decrease in rainfall and a remarkable temperature increase leaves very little space for the *Quercus* species to continue their existence. Moreover, insect pests and some fungi also attack these plants, damaging their leaves and acorns (Muhammad et al., 2023). For example, Afshan et al. (2022) and Muhammad et al. (2023) have published new powdery mildew reports on *Quercus* trees in Pakistan.

In response to a lesser amount of precipitation, the direct response in *Q. coccifera* is the reduction of water transport capacity. This takes place via reduction of vessel diameter without changing the amount of xylem tissue or foliage (Villar-Salvador et al., 1997). Changes regarding photosynthetic electron transport rates were found to be low in *Q. coccifera* under severe drought stress conditions, as an adaptation to xeric conditions, decreasing ETR photo-inactivating PSII centers (evidenced by their low predawn Fv/Fm ratios at high water stress, largely due to Fm quenching) (Peguero-Pina et al., 2009). The species shows extremely high resistance to drought-induced xylem cavitation (Vilagrosa, 2002). The plantations with closed canopy are less sensitive to moderate droughts, and more vulnerable to severe droughts.

Q. coccifera is a drought-resistant species (Ozturk et al., 2010a).

4.3.3 Other threats

Other threats include those specific to *Quercus.* The most important one is the livestock visiting oak forests and consuming the acorns, leading to a regeneration crisis (Rahman et al., 2022). *Q. coccifera* is especially preferred by goats, taken by goatherds who generally turn to oak forests for grazing their herds. When irregular and over-grazing continues for a long time, it causes serious damage to the forests, including *Q. coccifera* stands in Turkiye. For this reason, some precautions need to be taken for preventing damage due to overgrazing in the oak stands (Babalik et al., 2021).

Agricultural expansion and climate change are the other threats faced by the oak forests (Vinha et al., 2016). Population pollution, followed by the complete dependence of rural communities for fuel on different forests, including oak (oak is used as firewood, both for domestic as well as commercial purposes (Vinha et al., 2016)), creates an additional strain on local natural resources (Khan et al., 2010; Muhammad et al., 2023).

4.3.4 Climate change

Currently, one of the greatest challenges to humanity is climate change; it is of major international concern, directly threatening a large percentage of living beings. Some authorities say that global climate change is simply global warming, but the change is in fact an integrated system of several atmospheric phenomena and their products (Ozturk et al., 2010b, 2015,

2022; Imanberdieva et al., 2018; Ozturk & Altay, 2021).

Babalik et al. (2021) have reported that the potential distribution of *Q. coccifera* under bioclimatic conditions in 2050 and 2070 decreases in all model data scenarios. The most dramatic decrease is reported for 2070. These forecasts anticipate that the geographic spread of *Q. coccifera* will decrease in the future and potential habitat loss may reach serious levels in all climate-change scenarios (Babalik et al., 2021).

4.4 Conclusion

Oaks are long-living plants that have played a crucial role throughout the history of human civilizations (Martinez-Melendez et al., 2022). Looking at our earth's health and ecosystems, oaks appear as an important component. They have played a great role in human culture and spiritual practices. These plants have been evaluated for a wide range of goods and services. According to Wang et al. (2022) these forests need to be preserved as irreplaceable cultural heritage for future generations. Unfortunately, we as humans are continuously cutting and clearing these forests.

Muhammad et al. (2023) have reported that oak tree taxa play a critical role in the normal functioning of our ecosystems. They help to upkeep our biodiversity, help in the conservation of our water and soil resources, together with carbon sequestration. Our raw materials for timber, starch, tannin, cork and medicines come from these trees. Oak trees are essential for improving soil quality as well as maintaining healthy soil conditions (Rahman et al., 2022). They help in the prevention of soil erosion and preserve our water resources with the help of their roots and associated soil microbiomes. As mentioned by Wang et al. (2022), in addition to the multitude of benefits from oak forests, when leaves fall from these trees, or when these trees die, they decompose and fertilize the soil, allowing other plants to grow and thrive. Any decrease in the oak forests, even to a small extent, will detrimentally affect regional biodiversity, particularly when we are facing a challenge in the failure to regrow oak forests. This is a global issue now facing us (Dey, 2014).

References

AFSHAN N.U.S., ZAFAR I., JABEEN M., LIAQAT N. & KHALID A.N. 2022. *New records of powdery mildews on oak trees in Pakistan*. Nova Hedwigia: 79–88.

ALTAY V. 2009. *Istanbul'un Anadolu yakasi'nin kentsel vejetasyonu*. Ph.D. thesis. Marmara University, Istanbul.

ALTAY V, OZYIGIT II & YARCI C. 2012. *Plant communities in urban habitats of Istanbul-Turkey*. Pakistan Journal of Botany, 44: 177–186.

ALTAY V, SERIN M, YARCI C. & SEVEROGLU Z. 2012b. *Phytoecological and Phytosociological Investigations of the Vegetation of Gölcuk (Kocaeli/Turkey)*. Ekoloji, 21(84):74–89.

BABALIK A.A., SARIKAYA O. & ORUCU O.K. 2021. *The current and future compliance areas of Kermes Oak (Quercus coccifera L.) under climate change in Turkey*. Fresenius Environmental Bulletin, 30(1): 406–413.

DEY D.C. 2014. *Sustaining oak forests in eastern North America: Regeneration and recruitment, the pillars of sustainability*. For. Sci., 60: 926–942.

GUERRERO-SÁNCHEZ V.M., LÓPEZ-HIDALGO C., REY M.D., CASTILLEJO M.Á., JORRÍN-NOVO J.V. & ESCANDÓN M. 2022. *Multiomic Data Integration in the Analysis of Drought-Responsive Mechanisms in Quercus ilex Seedlings*. Plants, 11: 3067.

HAAVIK L.J., BILLINGS S.A., GULDIN J.M. & STEPHEN F.M. 2015. *Emergent insects, pathogens and drought shape changing patterns in oak decline in North America and Europe*. For. Ecol. Manag., 354: 190–205.

IMANBERDIEVA N., IMANKUL B., SEVEROĞLU Z., ALTAY V. & OZTURK M. 2018. *Potential impacts of climate change on plant diversity of Sary-Chelek Biosphere Reserve in Kyrgyzstan*. In D. Egamberdieva & M. Ozturk (eds), *Vegetation of Central Asia and Environs*, Springer Nature. Switzerland AG: 365–381.

KARAER F., KILINÇ M. & KUTBAY H.G. 1999. *The woody vegetation of the Kelkit Valley*. Turkish Journal of Botany, 23: 319–344.

KHAN N., AHMED M., WAHAB M. & AJAIB M. 2010. *Phytosociology, structure and physiochemical analysis of soil in Quercus baloot Griff, Forest District Chitral Pakistan*. Pak. J. Bot., 42, 2429–2441.

KUTBAY H.G., KARAER F. & KILINÇ M. 1999. *Orta Karadeniz bölgesinde bulunan Quercus L. ormanlarının fitososyolojik yapısı*. OMU Fen Edebiyat Fakültesi Fen Dergisi, 9–10: 1–17.

MARTÍNEZ-MELÉNDEZ N., RAMÍREZ-MARCIAL N., GARCÍA-FRANCO J.G., CACH-PÉREZ M.J. & MARTÍNEZ-ZURIMENDI P. 2022. *Importance of Quercus spp. for diversity and biomass of vascular epiphytes in a managed pine-oak forest in Southern Mexico*. For. Ecosyst., 9: 100034.

MUHAMMAD N., CASTILLEJO M.Á., REY M.-D. & JORRÍN-NOVO J.V. 2023. *An Overview of Oak Species in Pakistan: Past, Present, and Future Research Perspectives*. Forests, 14: 777.

NATALINI F., ALEJANO R., VÁZQUEZ-PIQUÉ J., CAÑELLAS I. & GEA-IZQUIERDO G. 2016. *The role of climate change in the widespread mortality of holm oak in open woodlands of Southwestern Spain*. Dendrochronologia, 38: 51–60.

NAVEH Z. 1998. *From biodiversity to ecodiversityholistic conservation of the biological and cultural diversity of Mediterranean landscapes*. In P.W. Rundel, G. Montenegro & F.M. Jakstic (eds), *Landscape disturbance and biodiversity in Mediterraneantype ecosystems. Ecological Studies*, 136. Springer Verlag. Berlin: 127–140.

OZTURK M. & ALTAY V. 2021. *Role of* Quercus coccifera *(=* Q. calliprinos*) in the light of climate change scenarios in the Mediterranean Basin*. Plant & Fungal Research, 4(2): 8–20.

OZTURK M., ALTAY V., & DIMEYEVA L.A. 2022. *Climate Change, Biodiversity and Ecology of Caspian Sea: An Overview*. Climate Change Alleviation for Sustainable Progression: 78–94.

OZTURK M., DOGAN Y., SAKCALI M.S., DOULIS A. & KARAM F. 2010. *Ecophysiological responses of some maquis (*Ceratonia siliqua L., Olea oleaster Hoffm. & Link, Pistacia lentiscus and Quercus coccifera L.*) plant species to drought in the east Mediterranean ecosystem*. Journal Environmental Biology, 31: 233–245.

OZTURK M., GUCEL S., KUCUK M. & SAKCALI S. 2010b. *Forest diversity, climate change and forest fires in the Mediterranean region of Turkey*. Journal of Environmental Biology, 31: 1–9.

OZTURK M., HAKEEM K.R., FARIDAH-HANUM I. & EFE R. 2015. *Climate Change Impacts on High-altitude Ecosystems*. Springer. Cham.

OZYIGIT S, ALTAY V, OZYIĞIT II & YARCI C 2015. *Vegetation ecology of the Princes' Islands, Istanbul-Turkey*. Journal of Enviromental Biology, 36: 113–120

PARASKEVOPOULOS S.P., IATROU G.D. & PANTIS J.D. 1994. *Plant growth strategies in evergreen-sclerophyllous shrublands (Maquis) in Central Greece*. Vegetatio, 115(2): 109–114.

PEGUERO-PINA J.J., SANCHO-KNAPIK D., MORALES F., FLEXAS J. & GIL-PELEGRIN E. 2009. *Differential photosynthetic performance and photoprotection mechanisms of three Mediterranean evergreen oaks under severe drought stress*. Functional Plant Biology, 36: 453–462.

PEREVOLOTSKY A. & HAIMOV Y. 1992. *The effect of thinning and goat browsing on the structure and development of Mediterranean woodland in Israel*. Forest Ecology and Management, 49: 61–74.

RAHMAN A., KHAN N., BRÄUNING A., ULLAH R. & RAHMAN I.U. 2022. *Effects of environmental and spatial gradients on Quercus-dominated Mountain forest communities in the Hindu-Kush ranges of Pakistan*. Saudi J. Biol. Sci., 29: 2867–2877.

SIMEONE M.C., GRIMM G.W., PAPINI A., VESSELLA F., CARDONI S., TORDONI E., PIREDDA R., FRANC A. & DENK T. 2016. *Plastome data reveal multiple geographic origins of Quercus Group Ilex*. Peer J, 4, e1897: 1–31.

TANTRAY Y.R., WANI M.S. & HUSSAIN A. 2017. *Genus* Quercus*: an overview*. International Journal of Advance Research in Science and Engineering, 6(8): 1880–1886.

TAVŞANOĞLU Ç. & COŞKUN U. 2009. *Köprülü Kanyon Milli Parkı'nda (Antalya) bulunan bazı maki türlerinin gelişme formu üzerinde keçi otlamasının etkisi*. Ekoloji, 18(72): 74–80.

TSIOURLIS G., KONSTANTINIDIS P. & XOFIS P. 2009. *Syntaxonomy and Synecology of Quercus coccifera Mediterranean Shrublands in Greece*. J. Plant Biol., 52: 433–447.

UGURLU E. & SENOL S.G. 2005. *Quercus-dominated vegetation units in the Aegean region of Turkey*. Botanica Chronica, 18: 283–291

VILAGROSA A. 2002. *Estrategias de resistencia al déficit hídrico en Pistacia lentiscus L. y Quercus coccifera L*. PhD Thesis; Departamento de Ecología, Universidad de Alicante; Alicante; Spain.

VILLAR-SALVADOR P., CASTRO-DIEZ P., PÉREZ-RONTOMÉ C. & MONTSERRAT-MARTI G. 1997. *Stem xylem features in three Quercus*

(Fagaceae) species along a climatic gradient in NE Spain. Trees, 12: 90–96.

VINHA A.F., COSTA A.S.G., BARREIRA J.C., PACHECO R. & OLIVEIRA M.B.P. 2016. *Chemical and antioxidant profiles of acorn tissues from Quercus spp.: Potential as new industrial raw materials*. Ind. Crops Prod., 94: 143–151.

WANG Y., XU C., WANG Q., JIANG Y. & QIN L. 2022. *Germplasm Resources of Oaks (Quercus L.) in China: Utilization and Prospects*. Biology, 12: 76.

WESTPHAL C., VON OHEIMB G., MEYER-GRÜNEFELD M., TREMER N., HÄRDTLE W., LEVANONY T., DAYAN T. & ASSMANNA T. 2009. *Ya'ar Bar'am-An old Quercus calliprinos forest of high nature conservation value in the Mediterranean region of Israel*. Israel Journal of Plant Sciences, 57: 13–23.

CHAPTER 5

SILVICULTURE AND RESTORATION OF *QUERCUS ILEX* WOODLANDS

ROBERTO MERCURIO

5.1 Distribution range and types

Species are widely distributed in the Mediterranean basin (Seigue 1985; Barbero et al., 1992; Terradas 1999; Quezel & Medail 2003). In Italy, holm oak forests occur between 0–200 metres up to 1,000 metres above sea level on the peninsula (1,200 metres in the Calabria region), and between 0–200 metres up to 1,000 metres on the two main islands (Sicily and Sardinia) (1,400–1,500 metres on Mount Etna in Sicily).

Fig. 5–1. Mixed stands of Fagus sylvatica *and* Quercus ilex *in Calabria region – Southern Italy (Photo Roberto Mercurio).*

Fig. 5–2. Coppices of holm oak with cork oak (Tyrrhenian side, Calabria region – Southern Italy (Photo Roberto Mercurio).

The European classification of the forest types (Barbati et al., 2006, 2014) includes the holm oak coenoses in the:

6.9.1 Mediterranean evergreen oak forest
Holm oak woodland
Relationship to Annex I of EEC 92/43 "Habitat" Directive
6310 *Dehesas* with evergreen *Quercus* spp,
6340 *Quercus ilex* and *Quercus rotundifolia* forests

Relationship to CORINE Biotopes
45.3 – Meso- and supra-Mediterranean holm-oak forests (*Quercion ilicis*)
Relationship to EUNIS Habitat Classification
G2.1 Mediterranean evergreen oak woodland

According to Bernetti (2015) the most important Italian holm oak forest types are:

- termophilous holm oak forests: in these coenoses with *Quercus ilex* live numerous sclerophyllous species as *Olea europea* var. *sylvestris*;
- typical holm oak forests (meso-Mediterranean): those occurring in the optimal climatic conditions;
- transition holm oak forests: *Quercus ilex* is mixed with sclerophyllous species and deciduous species such as *Fraxinus ornus, Quercus cerris, Acer monspessulanum, Acer campestre, Acer obtusatum, Ostrya carpinifolia, Sorbus domestica*;
- supra-Mediterranean holm oak forests: these coenoses are islands within the fringe of deciduous vegetation, where *Quercus pubescens, Quercus dalechampii, Quercus virgiliana* can become co-dominant species;
- rupicolous montane holm oak forests: relicts on rocks in the montane belt of the Apennines, typically dominated by *Fagus sylvatica*.

These are conservation nuclei of the species. In the Calabria region along the Jonian side it is also possible to list the holm oak forests mixed with the Hungarian oak (*Quercus frainetto*) (Cameriere et al. 2006; Mercurio & Spampinato, 2006; Mercurio et al., 2022).

Fig. 5–3. Holm oak forests with Hungarian oak. Jonian sector, Calabria region – Southern Italy (Photo Roberto Mercurio).

For the southern and insular Italian districts, Del Favero (2008) proposes the following forest types:

- primeval holm oak coenoses on cliffs;
- holm oak coenoses on calcareous substrates with maquis species;
- holm oak coenoses on calcareous substrates with European hop-hornbeam (*Ostrya carpinifolia*) and manna ash (*Fraxinus ornus*);
- holm oak coenoses on alterable calcareous substrates with species of the group of pubescent oak (*Quercus pubescens* s.l.);
- holm oak coenoses on siliceous substrates with cork oak and/or species of the heather moorland;
- typical holm oak coenoses of the basal and submontane belts;
- typical holm oak coenoses of the montane belt;
- holm oak coenoses of the montane belt with penetration of the European beech (*Fagus sylvatica*), or Turkey oak (*Quercus cerris*) and Hungarian oak (*Quercus frainetto*);
- coenoses of holm oak mixed with brooms species on volcanic soils.

Fig. 5–4. Penetration of holm oak in the montane belt. Monte Capo la Serra, southern Apennines, Campania region (Photo Rocco Carella).

Fig. 5–5. Penetration of holm oak in the montane belt. Monti di Bella, Muro e Avigliano, southern Apennines, Basilicata region (Photo Rocco Carella).

5.2 Mentions of reproductive biology

5.2.1 Gamic regeneration

Large seed production occurs starting from the age of 20 (6–8 in shoots), from 15–20 years according to La Mantia et al. (2003), with a frequency of 2 and 3–4 years (Seigue, 1985; Bernetti, 1995). Seed dispersal is by gravity (barochory), but could also be zoochore, and in this case greater distances can be achieved, up to 1 kilometre (Bossema, 1979; Cavalli et al., 2000; Gomez, 2003).

Dissemination occurs during winter/fall (from October to February). Not the whole seed production generates seedlings, since part of this is preyed on (both pre and post-dispersal) by insects, birds and mammals (rodents, wild boars). Furthermore, the environmental conditions of the stand affect seed germinability (Gomez, 2004). According to laboratory results, the germination capacity of seeds is high (80–90%) (Bartolini et al., 2001). Seed germination occurs in the spring following dispersal.

In Mediterranean environments shadow plays a fundamental role in the survival of young seedlings; indeed, the germination of holm oak seeds has more success in coppices, or in shadow stands, than in open fields (Bran et al., 1990; Pulido & Diaz, 2005). Concerning the blackberry (*Rubus* sp.) species presence, when this is not particularly abundant it is not a barrier but represents an advantage to holm oak regeneration. Similarly, Gomez-Aparicio et al. (2004) demonstrated the positive role played by shrubs on seedlings of different forest species (both in survival and in growth), including the holm oak. Seedlings predation (grazing, bark removal, poaching of soil) by domestic and wild herbivores can have an important role in the dynamics of Mediterranean forest coenoses, representing a threat to regeneration that can lead to deep changes to the ecosystems' function and the disappearance of the forest ecosystem. Others abiotic variables that can drive the young seedlings to death are fire passage, warming of the soil, extremes temperature (high and low) of the air, and primarily summer drought, in particular in first-year seedlings.

The seedlings density (of the first or a few years) sufficient to ensure regeneration in holm oak forests is $0.3/m^2$ (Susmel et al., 1976).

5.2.2 Agamic regeneration

Agamic regeneration occurs through stumps, roots or stems (Espelta et al., 1999). The duration of the dormant buds and then the resprouting capacity is 250 years in the holm oak (Serrada Hierro et al., 2004), and 200 years according to Seigue (1985) and Gambi (1986). However, it is influenced by carbohydrate availability, stump size, frequency and intensity of disturbance, intraspecific competition. Agamic regeneration allows the area damaged by disturbance or cuttings to recover quickly, ensuring soil protection. Giovannini et al. (1992) demonstrated that in a coppice dominated by holm oak, the area is reoccupied again by forest within three years of cutting.

5.3 Holm oak woodlands silviculture

5.3.1 Coppice system

This kind of silvicultural management fits excellently with the high capacity of the holm oak in shoots production following disturbance or cuttings even in old age, and for the assortments variety that can be obtained.

Coppicing with standards

The most used treatment is clearcutting with standards release (coppicing with standards – CWS). During cutting, all the shoots are removed while standards of 2–3 chronological classes are released.

In the past, in southern Italy, in addition to the cutting of shoots, *dicioccatura* (undercutting at the foot of stump with part of roots, used as well as firewood) was used frequently to increase the quantity of wood to remove at the end of every cycle (Marziliano, 2009).

Standards perform different functions: regeneration (to obtain plant born by seed that can progressively substitute the dying stumps), feeding (of wildlife fauna), production (of particular wood assortments), protection of soil and stumps, from assolation and wind. They also perform functions related to aesthetic value and biodiversity conservation. Furthermore, the presence of aged individuals can become useful for a conversion to the compound coppice.

Standards of good quality have a DBH (diameter at breast height) > 10–15 cm, straight stem rather than two-forked, crown well developed in the upper third, a low height–diameter ratio and are without diseases. Standards density has to be 50–60 per hectare (La Marca et al., 1998), while according to Del Favero (2008) it should not be more than 100 per hectare.

Fig. 5–6. Holm oak coppice with standards after cutting. Tuscany region – central Italy (Photo Roberto Mercurio).

The spatial arrangement of standards is generally uniform, with regular distances (Piussi & Alberti, 2015), but it can also be in strips and in groups (Bernetti et al., 2012) to enhance system complexity and mitigate impacts. Experience releasing groups of 20 individuals demonstrated the following advantages:

- higher stability and lower impacts related to strong plant isolation;
- creation of microhabitats (higher presence of ecotones, with positive consequences on biodiversity of the stand);
- higher regenerative capacity;
- higher ecological diversity in the clearings and a higher animal biodiversity within the groups;
- simplification of extraction operations with lower damages (Savini, 2010).

Rotation length in holm oak coppices was in the past 12–15 years; now it is 14–25 years (Fabbio & Cutini, 2017). Increased rotation length in holm oak coppices generally shows positive ecological and productive effects, allowing to obtain quality timber products without risks for agamic regeneration. Furthermore, lengthening rotation ensures the reduction of the harvesting surface, for an equivalent biomass amount, with positive consequences on the landscape (Hermanin & Pollini, 1990). Del Favero (2008) suggests distinguishing the coppices rotation on the different forest types: Fifteen years in the typical holm oak forests, 25–30 years in the holm oak coenoses with *Fraxinus ornus, Ostrya carpinifolia* and *Quercus pubescens* s.l., and > 30 years in holm oak forests with sclerophyllous maquis species or with Cork oak.

Recently, the method of coppicing with standards release has been criticized since it causes serious impacts on environment and landscape. For this reason, mitigation measures, further to the customary (traditional) techniques, such as the interruption of grazing in the first 3–4 years following the harvesting, are suggested. These are:

- the release of forest strips along roads, ridges, rivers, stems, and impluvium;
- the release of sporadic forest species occurring in the coenoses;
- the preservation of veteran trees;
- the release of harvesting residuals in rows, in particular along slopes, as well as where there is no risk of fires and of pathologies;
- the prohibition on the complete use of epigeal biomass (full-tree logging), at least on the slopes and in those stands exposed to erosion risk;
- the adoption of concentration and extraction measures and management of low-impact forest viability solutions.

The size of the cutting area is related to the economic and ecological sustainability of interventions. Indeed, it influences the timber mass amount harvestable, worksite organization and impacts on environment and landscape. To reduce the environmental impact, the cutting surface should be less than 10 hectares in roughly flat stands, with a further reduction for stands with an incline over 30%. The time interval between two successive cuttings should not be less than three years. The shape of the cutting area has to be larger than longer for aesthetic and ecological considerations; another solution can be represented by a cutting area interposed between standard release in strips, or between wood strips not affected by silvicultural interventions along the height profiles (Del Favero, 2008). Cutting width and shape are also related to the impact of damage by wild ungulates (Giovannini et al., 2003); cutting one area of 5 hectares width is better than cutting five areas of 1 hectare width, since by reducing the ecotonal areas the stands become less attractive to ungulates.

In sites with high environmental sensitivity, it can be useful to avoid uniform forest management, alternating coppices with converting coppice into high forest where it is possible, as well as stands left to their natural evolution. This requires appropriate planning to evaluate the effects of the different silvicultural interventions on the landscape. In this respect, a study on visuals from the big communication roads (Scarfò et al., 2010) represents a novel approach to reduce the aesthetic impact of cuttings in the first years after the intervention.

Fig. 5–7. Wide coppicing in pure holm oak forests in Calabria region – southern Italy (Photo Roberto Mercurio).

Coppice selection system

Less used is the coppice selection system (CSS; *taglio a sterzo, taglio della formica* in Italian). It is traditionally adopted in Tuscany (central Italy) and in similar forms in other Mediterranean countries (Seigue, 1985; Montoya, 1989; Retana et al., 1992; Mayo & Rodà, 1993; Gracia, 1998).

During cutting, part of the shoots and standards are removed. The stand is composed of two layers, the lower with shoots of different age classes, the higher with the standards. According to the method adopted in the Montagnola Senese (Tuscany) (Gambi, 1986) the standards number after cutting was 800–1,000 per hectare (with age classes multiples of the felling cycle). This means the interval time between cuttings is seven years, while the felling diameter is 8 centimetres.

Compound coppice

This treatment, traditionally applied in Tuscany, could be listed among sylvo-pastoral systems, since it is used both for timber production and as a typical method of wild or semi-wild husbandry (Bernetti, 1987; Fabbio, 2009). The structure is composed of a dominated layer with even aged shoots (coppice component), and an overstory layer (high forest component) of different ages (generally more than three classes) that can be chosen among holm oak individuals with narrow crown, and eventually among cork oak individuals (Bernetti et al., 2012). In the typical form, the standards number of different class ages was 140–160 per hectare. La Marca (2006), as a consequence of analyses made on holm oak coppices (Gargano promontory, Apulia region – southern Italy), demonstrates that a compound coppice with 140 standards per hectare gives a high production of fruits for livestock and wild fauna feeding, good mechanical stability of the individuals and also significant timber production; lastly compared with coppice with standards release the compound coppice shows in addition positive effects on environment and landscape.

For old mixed compound coppices with pubescent oak and holm oak in the Chianti area (Tuscany), Becagli et al. (2006) propose a progressive rejuvenation of the high forest component, avoiding creating excessive lacks in forest cover.

These last two treatments, the coppice selection system and the compound coppice, due to their evident cultural values, should be conserved to preserve important traditional knowledge and skills.

Fig. 5–8. Traditional compound coppice at Montagnola Senese, Tuscany (Photo Roberto Mercurio).

Truffles and grazing are other important resources of the holm oak forests. *Quercus ilex* is an interesting species for truffle farming, since it is a symbiont of the black truffle (*Tuber melanosporum* Vitt) (Bencivenga & Vignozzi, 1989). Diette & Lauriac (2009) propose a "truffle silviculture" to increase truffle production, acting in particular on silvicultural interventions of thinnings and cleanings. Particular attention must be paid to coppicing, since in coppices of 20–25 years the black truffle production stops 4–6 years after cutting; treatments should be done in March or November to favour truffle production.

Holm oak woodlands fit the wild grazing of pigs, both for the quality of acorns and for their seasonality, which cover 5–6 months in the year. Fabbio (2009) reports acorn production per tree from 20–30 kilograms to 150–2,000 kilograms, while the acorn production for a holm oak high forests of high fertility in four years can vary from 426 to 3,253 kilograms per hectare. The average production in the Spanish *dehesas* (density of 50 trees per hectare) is 250–600 kilograms per hectare (Gea-Izquierdo et al., 2006). Dense forests in France, with 1,400 trees per hectare, give an average acorn production of 512 kilograms per hectare each year.

5.3.2 Conversion to high forest

Rural land abandonment and public criticism of ecological and aesthetic effects have forced public administrations in the last decades to privilege the conversion of coppices in more complex forest systems. Holm oak high forests are justified where protective, aesthetic, recreational functions must be assolved, or when grazing is needed.

Conversion of coppice with standards in compound coppice

Method of progressive standard releases (Del Favero, 2008).

At the end of the rotation period, coppice is cut and a certain number of standards of the first cycle and standards of the other cycles are released to drive the number and distribution of the different chronological classes to that fixed for the compound coppice and to reach a balance between coppice and high forest (Ciancio & Nocentini, 2004). Standards must be chosen among those trees with the best phenotypic characteristics of the species, both of the dominant species that of the occurring compaign species uniformly distributed (Ciancio et al., 1983).

Conversion of coppice forests into high forest

Several authors (Ducrey, 1992; Ducrey & Toth, 1992; La Marca et al., 1995; Cutini & Mascia, 1998; Fabbio et al., 1998; Serrada et al., 1998; Dagis & Kakouros, 2006; Bravo, 2008; Manetti et al., 2013) agree that the interruption of coppicing, associated with conversion to high forest interventions, drives secondary successions characterized by higher functionality, complexity and biological richness of the coenoses. Studies conducted by La Marca et al. (2008) have shown that the volume increments recorded in the stands involved in conversion into high forest interventions are significantly higher than in stands left to natural succession after coppicing interruption. Cotillas et al. (2009) demonstrated that thinnings in holm oak coppices enhance water availability in the Mediterranean environment.

Fig. 5–9. Aged holm oak coppices after thinnings interventions. Tuscany (Photo Roberto Mercurio).

Different methods of conversion into high forest have been implemented.

Conversion by ageing. Bernetti (1995) recommends modifying the original method – a series of thinnings from below; that removing the fuel biomass also plays a role in fire prevention, and favours crown enlargement. At 85–90 years when the density is 600–700 trees per hectare, the actual conversion cut (seed cut) is made that will lead to a monolayer high forest.

Coppice conversion cuts. Amorini et al. (1998) applied this method on a 45-year-old coppice in Sardinia, with a mixed thinning from below with intensity of 50–60% of the basal area to accelerate evolution trend towards structures more complex and mature, without affecting the microclimatic conditions of the stand or inducing disturbance (Cutini & Mascia, 1998; Fabbio et al., 1998).

Intensive standard density (selection and intensive management of crop trees). Experiments carried out in Aspromonte (Calabria region – southern Italy) on 30-year-old holm oak coppices have compared:

- low intensity interventions repeated with short intervals;
- release of shoots >3–3.5 metres height on dominated stumps and of 1–2 shoots among the best phenotypes on the dominant stumps.

The first results have demonstrated that in the first case, removing 30% of the biomass (about 60% of the total shoots), stand structure and soil conservation conditions are not significantly affected and, at the same time, there are no recorded impacts on the environment and landscape (Ciancio et al., 2001; Iovino & Menguzzato, 2001).

Fig. 5–10. First conversion cut with the intensive standard density method. Aspromonte (Calabria region) (Photo Roberto Mercurio).

Progressive standard density method. In this case, at the time of coppicing more than 100 standards of the first rotation period, and also 20–25 standards of the previous rotations, are released, thus turning to compound coppice (Del Favero, 2008).

5.3.3 The aged coppices issue

The abandonment of coppicing recorded in Italy in recent decades has meant that 45% of holm oak coppices are more than 40 years old (Fabbio & Cutini, 2017), opening the discussion on future management, with the following options:

Recovery of coppicing. This is compatible from a biological point of view due to the high resprouting capacity of the stumps, as well in old age (50 years old) (Bernetti, 2015; Fabbio & Cutini, 2015), and also from the perspective of relaunching the local economy and short supply chains. According to Fabbio & Cutini (2017) holm oak coppices located along the coastal environment of Sardinia show a culmination of mean volume increment in the interval of 40–60 years and in two cases no culmination between 65 and 70 years. Sferlazza et al. (2018), following analysis on holm oak coppices in Sicily, argue that coppicing can be considered a sustainable forest management technique from the point of view of the carbon balance, since biomass is used for energy purposes. Another option experimented in Spain is to carry out thinnings from below in abandoned or even aged coppices, removing 30–50% of trees every 18–25 years, to shift towards uneven aged structures (Mayor & Rodà, 1993).

Coppices in silviculturally oriented evolution toward high forests. This option can be applied in the best forest stands where aesthetic and recreational functions are preferred.

Coppices left to natural evolution. In this case, the forest portion is conserved with natural ageing, allowing a higher ecological coherence to be realized, enhancing the biodiversity values and creating a more complex mosaic.

Coppices managed for special purposes. In this case interventions are suspended to achieve ecosystem services, such as CO_2 fixation, biodiversity and soil conservation, enhancing of water quality – functions that become more important than coppicing. Some studies provide fundamental suggestions as to the best choices for a better strategy. In aged (55-year-old) holm oak coppices in Gargano (Apulia), standing volume biomass is 287 m^3 ha^{-1} , the total dead volume biomass is 33.89 m^3 ha^{-1}, while the stocked carbon is 12.62 megagrams (Mg) of carbon (C) per hectare $(ha)^{-1}$ (Marziliano, 2009). These data clearly show that the death biomass quantity could trigger phytosanitary problems, and in particular in the Mediterranean region enhance fire risk, since standing dead wood increases the propagation and intensity of fire.

The different management options described can be harmonized and implemented to enhance landscape and ecological complexity (Mairota et al., 2016).

A different aspect is represented by coppice management in the protected area (as it can be in those Sites of Community Interest (SCI) or Special Protection Areas (SPA) of the Natura 2000 Network, where the 9340 code of Annex I of the Habitat Directive is found), and in these cases it is better to arrest the coppicing, or to continue it with the purpose of maintaining local traditional management. At the current stage, there are no studies demonstrating that abandonment of coppicing may represent the best solution to achieve the conservation of holm oak habitats (Mairota et al., 2016).

Fig. 5–11. Aged holm oak coppices (Chianti, Tuscany region, central Italy) (Photo Roberto Mercurio).

Fig. 5–12. Firewood obtained by aged holm oak coppices in Calabria (Photo Roberto Mercurio).

5.3.4 High forest management

Monolayer, even aged holm oak high forests are not widespread and are not suitable for high wood quality production. Up to the present day, the different management proposals from different authors are still not supported by scientific evidence. Gambi (1986) proposed a shelterwood system, considering also regeneration cutting coppices in conversion, with the following sequence: low–medium intensity seed cutting when the stand has 400–500 trees per hectare, final removal cut after about ten years, in a rotation of 100–120 (150–200) years.

In Italy, in particular in the southern peninsula and on the main islands (Sicily and Sardinia), there are more widespread uneven-aged holm oak high forests, in stands used for pigs grazing where dead trees were removed and interested by selection cutting of decaying individuals that were not anymore able to produce acorns. This management method could now become useful for the wild grazing of pigs, representing a valid alternative to intensive pig breeding, which is highly dangerous for the environment and public health.

5.3.5 Restoration of degraded holm oak stands

A degraded forest stand shows a "reduction of the capacity of a forest to provide goods and services" (FAO; 2011; Lund, 2014). Holm oak stands are highly resilient (Gracia & Retana, 2004) having a strong tolerance to dryness in terms of photosyntetic potential and efficiency in water utilization (Di Matteo et al., 2014). The deepening of the root system allows to maximize water absorption (Barbeta et al., 2015; David et al., 2016), while holm oak shoots react well to the wild ungulates and livestock grazing. However, the repeated and combined effects of fire, grazing and intense cuttings with short intervals, have severe impacts as well on holm oak stands (Soler & Sala, 1992).

Forest restoration must be based on models, otherwise it is not possible to monitor and evaluate the effectiveness of each intervention. For this purpose, some old growth forests since utilized for grazing are very important, such as 'A Mancusa (Calabria) (Mercurio et al., 2022),

Supramonte (Sardinia) (Susmel et al., 1976), and also in Corsica (Panaiotis et al., 1998).

Fig. 5–13. 'A Mancusa forest (Aspromonte, Calabria, southern Italy) (Photo Roberto Mercurio).

Post-fire restoration

Holm oak has a strong capacity to resprout after exposure to fire (Gracia & Fretana, 2004; Broncano et al., 2005), while it shows a low capacity for seed dispersion over long distances (Gracia et al., 2002).

The post-fire recovery of the holm oak stands is also strong when all the above-ground biomass is totally removed by fire, since it is rapidly replaced; thus, the interventions needed are moderate, consisting just of facilitating and accelerating the process of restoration. In their studies Jacquet & Prodon (2007) show that 40 years after fire are needed for the recovery of a holm oak stand, while for the associated bird community 30 years are sufficient.

Generally, a single disturbance event does not damage the stand since it is capable of repairing itself quickly. Ferran & Vallejo (1998) demonstrated that, five years after fire, the full coverage of the burned area by forest is again recorded. However, repeated events or intense fire causes a deep degradation, with the appearance of herbaceous species of acid soils, subsequently shrubs such as *Cistus* sp. and *Calicotome* sp. Finally, if no further disturbance factors appear, this stand can shift towards an acidophilus wood dominated by the cork oak (*Quercus suber*); once this process is accomplished it becomes complicated to recover the holm oak coenoses, and this can be done just through targeted active restoration interventions.

Low density stand cases

Experiments conducted by Sardin et al. (2003) in southern France demonstrated that plantation using containers gives better results in terms of recovery and height increment than direct seeding, although it is more expensive. In post-fire interventions in Tuscany, consisting of plantation of *Quercus ilex* with *Pinus pinea* and *Pinus halepensis*, the Aleppo pine showed a height increment five time greater than the holm oak. However, holm oak had a strong ability to hold the soil for its huge and deep taproot system, giving a greater guarantee of resprouting after fire; it is also less flammable compared with Mediterranean pines. More recent forest restoration techniques tend to exclude the pine species, using it just on eroded soils and on shallow soils, adopting innovative and precise seeding of holm oak. Resprouting Mediterranean broadleaved species are used in pinewoods to enhance their resilience (Grandos et al., 2016).

Generally, the replanting of trees and shrubs in stands occupied by Mediterranean maquis is not necessary; rather the "island" seeding of holm oak and cork oak seedlings irregularly distributed can be done just in those parts of the coenoses where, as a consequence of the intensity of the fire, maquis hesitate to resprout. Another solution in this case can be represented by hydroseeding with the local maquis species (Leone, 2001).

Plantation density is 600–1,000 individuals per hectare (3x3/4x4 metres), while in the plantation mycorrhized with truffles it is 5x5 till 7x7 metres

(200–400 individuals per hectare) (Villar-Salvador et al., 2013). Stones and mulching can be useful to reduce the drought effects on the holm oak seedling (Jimenez et al., 2016). Other solutions can be used to reduce soil erosion, such as seeding of herbaceous species, transverse barriers, mulching, etc. (Bautista et al., 2009; Vallejo et al., 2012; Prats et al., 2014; Pizarro-Tobias et al., 2015).

Following the recent destruction of the coastal pine forests of stone pine (*Pinus pinea* L.) in the Lazio region (central Italy) due to fires, *Tomicus* sp. and *Toumeyella parvicornis* outbreaks, secondary agents of the climate drying, the holm oak will be the main species to replace the pine forests in the more xeric parts of the sandy coasts (Mercurio, 2020, 2021).

Fig. 5–14. Pinewood of Castel Fusano (Rome) in the transformation process towards a holm oak forest (Photo Roberto Mercurio).

High density stand cases

In those cases where sufficient moisture remains in the stumps, the coppicing of burned stands is very effective. Indeed, after the appearance of annual species, soon the forest species start to resprout with vigorous shoots; the former are strawberry tree (*Arbutus unedo*) and tree heat *(Erica arborea*) followed by the holm oak.

Holm oak is favourable to salvage felling below ground (*tramarratura* in Italian) (Gambi & Proietti-Placidi, 1991) to reactivate stump vitality and assist dormant bud emission. To solicit the holm oak stumps in the emission of new root suckers, a cutting known as *débourdage* or *saut du piquet* (in French) can be applied. This consists of removing shoots at 20–25 centimetres from the ground and then hitting them with the back of a wood chopping axe used for splitting stumps (Coste, 1984; Gambi, 1986). To prevent new disturbance events in the season following the fire all the dead shoots are felled; this wood material will be used by the local population as firewood. Moreover, this intervention removes the blackened image of the burned forest. By adopting mechanical brush cutters, it is possible to chip the burned vegetation, leaving the residuals uniformly on the soil. This technique enriches the soil in organic matter, improving its physico-chemical and biological properties.

Manual cutting is limited to those areas where the surface incline, morphology and high stone content do not allow mechanical means to be used.

According to Puxeddu & Delogu (2007) cutting interventions in the maquis after fire have not proved effective, since plots where cuttings, and following selection cuttings of the resprouting shoots, have been conducted have not recorded higher biometrical parameters.

In stands with an incline over 40%, the material after cutting can be assembled in bundles arranged along the height profiles (Hofmann, 2008).

In the specific case of the reconstitution of holm oak forests, in addition to the described cultural practices, also effective is the pruning of decaying individuals (Trabaud, 1996), consisting of shortening the main branches at 30–40 centimetres from their insertion in the stem (pollarding) to facilitate vegetation regrowth.

References

AMORINI E., BRUSCHINI S., CUTINI A., FABBIO G. & MANETTI MC. 1996. *Silvicultural treatment of holm oak (Quercus ilex L.) coppices in Southern Sardinia: thinning and related effects on stand structure and canopy cover*. Annali Istituto Sperimentale per la Selvicoltura, 27: 167–176.

ANGELINI P., CASELLA L., GRIGNETTI A. & GENOVESI P. (eds). 2016. *Manuali per il monitoraggio di specie e habitat di interesse comunitario (Direttiva 92/43/CEE) in Italia: habitat*. ISPRA, Serie Manuali e linee guida, 142/2016.

BARBATI A., CORONA P. & MARCHETTI M. 2006. *European forest types. Categories and types for sustainable forest management and reporting*. European Environment Agency, EEA Technical report No. 9/2006.

BARBATI A., MARCHETTI M., CHIRICI G. & CORONA P. 2014. *European Forest Type and Forest Europe SFM indicators: tools for monitoring progress on forest biodiversity conservation*. Forest Ecology and Management, 321: 145–157.

BARBERO M., LOISEL R., QUÉZEL P. 1992. *Biogeography, ecology and history of Mediterranean* Quercus ilex *ecosystems*. Vegetatio, 99–100: 19–34.

BARBETA A., MEJÍA-CHANG M., OGAYA R., VOLTAS J., DAWSON T.E. & PEÑUELAS J. 2015. *The combined effects of a long-term experimental drought and an extreme drought on the use of plant-water sources in a Mediterranean forest*. Global Changing Biology, 21: 1213–1225.

BARTOLINI G., BUSSOTTI F., ASENSIO A., GARCÍA C., CHESSA I., CICCARESE C., CICCARESE L., CROSTI R., CULLUM F.J., DI NOI A., GARCÍA-FAYOS P., LAMBARDI M., LISC MI., LUCCI S., MELINI S., MUÑOZ EINOSO J.C., MURRANCA S., NIEDDU G., PACINI E., PAGNI G., PATUMI M., PÉREZ GARCÍA F., PICCINI C., ROSSETTO C. & TRANNE G. 2001. *Schede informative sulla propagazione per seme degli alberi e degli arbusti più diffusi della flora mediterranea*. In B. Piotto & A. Di Noi (eds), *Propagazione per seme di alberi e arbusti della flora mediterranea*. ANPA. Roma. pp. 143.

BAUTISTA S., ROBICHAUD P.R. & BLADÉ C. 2009. *Post-fire mulching*. In A. Cerdà & P.R. Robichaud (eds), *Fire Effects on Soils and*

Restoration Strategies. Science Publishers. Enfield. pp. 353–372.

BECAGLI C., CANTIANI P. & FABBIO G. 2006. *Trattamento sperimentale in un ceduo composto di roverella e leccio del Chianti senese: primi risultati*. Annali C.R.A. Istituto Sperimentale per la Selvicoltura, 33 (2002–2004): 31–38.

BENCIVENGA M. & VIGNOZZI G. 1989. *I tartufi in Toscana*. Il Vantaggio. Firenze. pp. 209.

BERNETTI G. 2015. *Le piante del bosco. Forme, vita e gestione*. Compagnia delle foreste. Arezzo. pp. 350.

BERNETTI G. 1995. *Selvicoltura speciale*. UTET.Torino. pp. 415.

BERNETTI G. 1987. *I boschi della Toscana*. Edagricole. Bologna. pp. 177.

BERNETTI G., DEL FAVERO R. & PIVIDORI M. 2012. *Selvicoltura produttiva*. Edagricole. Bologna. pp. 225.

BOSSEMA L. 1979. *Jays and oaks: an eco-ethological study of a symbiosis*. Behaviour, 70: 1–117.

BRONCANO M.J., RETANA J. & RODRIGO A. 2005. *Predicting the recovery of* Pinus halepensis *and* Quercus ilex *forests after a large wildfire in northeastern Spain*. Plant Ecology, 180(1): 47–56.

BRAN D., LOBREAUX O., MAISTRE M., PERRET P. & ROMANE F. 1990. Germination of *Quercus ilex* and Q. *pubescens* in a *Q. ilex* coppice. Long-term consequences. Vegetatio, 87: 45–50.

BRAVO J. A. 2008. *Selvicultura en montes bajos y medio de* Quercus ilex *L.,* Q.pyrenaica *Willd. y Q. faginea Lam*. In R. Serrada, G. Montero & J.A. Reque (eds), *Compendio de Selvicultura Aplicada en España*. INIA. Madrid, pp.657–744.

CAMERIERE P., GUGLIOTTA O.I., MERCURIO R., MODICA G. & SPAMPINATO G. 2003. *Primo inquadramento tipologico dei boschi di farnetto (*Quercus frainetto *Ten.) dell' Aspromonte (Italia meridionale)*. Monti e Boschi, 54(3–4): 3–8.

CAVALLI S., IOALÈ P., MACCHIA M. & SBRAGIA M. 2000. *Contributo sulla disseminazione zoocora di Quercus ilex in aree litorali della Toscana nord-occidentale*. Monti e Boschi, 51(1): 13–16.

CIANCIO O., IOVINO F. & MENGUZZATO G. 2002. *Prove sperimentali di avviamento a fustaia con il metodo del rillascio intensivo di allievi: i cedui di leccio in Aspromonte (Calabria)*. In O. Ciancio & S. Nocentini (eds), *Il Bosco ceduo in Italia*. Accademia Italiana di

Scienze Forestali. Firenze, pp. 325–342.

CIANCIO O., MERCURIO R., NOCENTINI S. & ECCHER A. 1983. *Tecniche di miglioramento e metodi di conversione e trasformazione*. In *La gestione dei cedui*. L'Italia Agricola, 120(4): 77–86.

CIANCIO O. & NOCENTINI S. 2004. *Il bosco ceduo*. Accademia Italiana di Scienze Forestali. Firenze. pp. 721.

COTILLAS M., SABATE S., GRACIA C. & ESPELTA J.M. 2009. *Growth response of mixed Mediterranean oak coppices to rainfall reduction. Could selective thinning have any influence on it?* Forest Ecology and Management, 258: 1677–1683.

CUTINI A. & MASCIA V. 1998. *Silvicultural treatment of holm oak (*Quercus ilex *L.) coppices in Southern Sardinia: effects of thinning on water potential, transpiration and stomatal conduttance*. Annali Istituto Sperimentale per la Selvicoltura, 27 (1996): 47–53.

COSTE D. 1984. *Une technique oubliée d'exploitation du chêne vert*. Forêt Méditerranéenne, 6(2): 167–169.

DAFIS S. & KAKOUROS P. 2006. *Guidelines for the rehabilitation of degraded oak forests*. Greek Biotope/Wetland Centre. Therni. pp. 40.

DAVID T.S., PINTO C.A., NADEZHDINA N. & DAVID J.S. 2016. *Water and forests in the Mediterranean hot climate zone: a review based on a hydraulic interpretation of tree functioning*. Forest Systems, 25(2), eR02. http://dx.doi.org/10.5424/fs/2016252–08899.

DEL FAVERO R. 2008. *I boschi delle regioni meridionali e insulari d'Italia*. Cleup. Padova. pp. 466.

DIETTE S. & LAURIAC A. 2006. *La sylviculture truffière: aperçu historiques, apports techniques et enjeux pour la région méditerranéenne*. Revue Forestière Française, 56(3): 219–230.

DI MATTEO G., PERINI L., ATZORI P., DE ANGELIS P., MEI T., BERTINI G., FABBIO G. & SCARASCIA MUGNOZZA G. 2014. *Changes in foliar carbon isotope composition and seasonal stomatal conductance reveal adaptive traits in Mediterranean coppices affected by drought*. Journal of Forestry Research, 25(4): 839–845.

DUCREY M. 1992. *Quelle silviculture et quel avenir pur les taillis de chene vert (*Quercus ilex *L.) de la région méditerranéene française*. Revue Forestière Française, 44(1): 12–33.

DUCREY M. & TOTH J. 1992. *Effect of cleaning and thinning on heigh growth and girth increment in holm oak coppices (*Quercus ilex *L.)*. In F. Romane & J. Terradas (eds), Quercus ilex *L. ecosystems: function, dynamics and management*. Kluwer Academic Publ. Dordrecht, pp. 365–376.

ESPELTA J.M., SABATÉ S. & RETANA J. 1999. *Resprouting Dinamics*. In F. Rodà, J. Retana, C.A. Gracia & J Bellot (eds), *Ecology of Mediterranean Evergreen Oak Forests*. Spinger, Berlin, Heidelberg, pp. 61–74. https://doi.org/10.1007/978-3-642-58618-7_5

FABBIO G. (ed.) 2009. *Il bosco e l'allevamento della razza Cinta Senese.* ARSIA. Regione Toscana, pp. 107–120

FABBIO G. & CUTINI A. 2017. *Il ceduo oggi: quale gestione oltre le definizioni?* Forest@, 14: 257–274.

FABBIO G., CUTINI A. & MASCIA V. 1996. *Silvicultural treatment of holm oak (*Quercus ilex *L.) coppices in Southern Sardinia: effects of canopy and crop thinning on microclimate*. Annali Istituto Sperimentale per la Selvicoltura, 27: 55–63.

FAO 2011. *Assessing forest degradation. Towards the development of globally applicable guidelines*. Forest Resources Assessment Working Paper 177, Rome.

FERRAN A. & VALLEJO R.V. 1998. *Long term plant regeneration after wildfires in Mediterranean ecosystems of NE Spain*. In L. Trabaud (ed.), *Fire Management and Landscape Ecology*. International Association of Wildland Fire. Fairfield Washington, pp. 155–166.

GAMBI G. 1986. *Le leccete*. Monti e Boschi (Le schede), 37(1): I–XII.

GAMBI G. & PROIETTI PLACIDI A.M. 1991. *Effetti della tramarratura sulla rinnovazione di un ceduo di querce*. Monti e Boschi, 42(5): 23–29.

GEA-IZQUIERDO G., CAÑELLAS I. & MONTERO G. 2006. *Acorn production in Spanish holm oak woodlands*. Investigacion Agrarias Sistemas y Recursos Forestales, 15(3): 339–354.

GIOVANNINI G., CHINES A. & GANDOLFO G. 2003. *Danni da ungulati selvatici in boschi cedui: effetti delle modalità di utilizzazione forestale*. Sherwood, 9(1): 9–16.

GIOVANNINI G., PERULLI D., PIUSSI P. & SALBITANO E. 1992. *Ecology of vegetative regeneration after coppicing in macchia stands in central Italy*. Vegetatio, 99/100: 331–343.

GÓMEZ J.M. 2004. *Importance of microhabitat and acorn burial on* Quercus ilex *early recruitment: non-additive effects on multiple demographic processes*. Plant Ecology, 172: 287–297.

GÓMEZ J.M. 2003. *Spatial patterns in long-distance dispersal of* Quercus ilex *acorns by jays in a heterogeneous landscape*. Ecography, 26: 573–584.

GÓMEZ-APARICIO L., ZAMORA R., GÓMEZ J.M., HÓDAR J.A., CASTRO J. & BARAZA E. 2004. *Applying plant facilitation to forest restoration in Mediterranean ecosystems: a meta-analysis of the use of shrubs as nurse plants*. Ecological Applications, 14: 1128–1138.

GRACIA M. 1998. *Les taillis de chêne vert, irregulières et furetées du nord-est de peninsule iberique*. Revue Forestière Française, 50(5): 467–478.

GRACIA M. & RETANA J. 2004. *Effect of site quality and shading on sprouting patterns of holm oak coppices*. Forest Ecology and Management, 188(1–3): 39–49.

GRACIA M., RETANA J. & ROIG P. 2002. *Mid-term successional patterns after fire of mixed pine-oak forests in NE Spain*. Acta Oecologica, 23: 405–411.

GRANADOS M.E., VILAGROSA A., CHIRINO E. & VALLEJO V.R. 2016. *Reforestation with resprouter species to increase diversity and resilience in Mediterranean pine forests*. Forest Ecology and Management, 362: 231–240.

HERMANIN L. & POLLINI M. 1990. *Produzione legnosa e paesaggio: considerazioni sul turno in cedui di leccio*. Cellulosa e Carta, 11(2): 6–10.

HOFMANN A. 2008. *Ricostituzione per via artificiale.* In S. Baldini (ed.), *Tecniche di ripristino dei boschi percorsi da incendio*. A.G.R.A. Roma, pp. 87–107.

IOVINO F. & MENGUZZATO G. 2001. *Valorizzazione colturale dei boschi cedui dell'Italia meridionale*. L'Italia Forestale e Montana, 56(5): 362–376.

JACQUET K. & PRODON R. 2007. *Résilience comparée des peuplements de chêne vert et de chêne liège après incendies*. Revue Forestière Française, 59(1): 31–44.

JIMÉNEZ M.N., FERNÁNDEZ-ONDOÑO E., RIPOLL M.Á., CASTRO-RODRÍGUEZ J., HUNTSINGER L. & NAVARRO F.B. 2016. *Stones and organic mulches improve the* Quercus ilex *L. afforestation success under Mediterranean climatic conditions*. Land Degradation and Development, 27: 357–365.

LA MANTIA T., CULLOTTA S. & GARFÌ G. 2003. *Phenology and growth of* Quercus ilex *L. in different environmental conditions in Sicily (Italy)*. Ecologia Mediterranea, 29(1): 15–25.

LA MARCA O. 2006. *Ricerche sperimentali per una gestione sostenibile dei cedui di leccio: opzioni colturali a confronto*. In J. Parrotta, M. Agnoletti & E. Johann (eds), *International Conference on Cultural Heritage and Sustainable Forest Management: the role of traditional knowledge*, 8–11 June 2006, Florence, Italy, pp. 138–144.

LA MARCA O., MARZILIANO P., MORETTI N., PIGNATTI G. 1995. *Dinamica strutturale e floristica in un ceduo di leccio*. Annali Accademia Italiana di Scienze Forestali, 44: 235–256.

LA MARCA O., MARZILIANO P. & SANTORO A. 1998. *Management in ageing holm oak (*Quercus ilex *L.) in coppices: experimental results after four growing seasons*. Annali Istituto Sperimentale per la Selvicoltura, 27 (1996): 161–166.

LA MARCA, O. MARZILIANO P.A. & SCOPIGNO D. 2008. *Opzioni selvicolturali su cedui di leccio del Gargano: risultati a 14 anni dall'avvio della sperimentazione*. Forest@, 5: 318–336.

LEONE V. 2001. *Interventi selvicolturali per il recupero di soprassuoli boschivi percorsi da incendi*. L'Italia Forestale e Montana, 56(6): 430–440.

LOSSAINT P. & RAPP M. 1978. *La forêt méditerranéenne de chênes verts*. In M. Lamotte & M. Bourliere (eds), *Problèmes d'écologie. Structure et fonctionnement des écosystémes terrestres*, Masson, Paris, pp. 129–182.

LUND H.G. 2014. *Definitions of Forest, Deforestation, Reforestation and Afforestation*. Online Report. Forest Information Services, Gainesville, Virginia, USA. http://home.comcast.net/%7Egyde/DEFpaper.htm.

MAIROTA P., BUCKLEY P., SUCHOMEL C., HEINSOO K., VERHEYEN K., HÉDL R., TERZUOLO P.G., SINDACO R. & CARPANELLI A. 2016. *Integrating conservation objectives into forest management: coppice management and forest habitats in Natura 2000 sites*. iForest – Biogeosciences and Forestry, 9(4): 560–568.

MAIROTA P., MANETTI M.C., AMORINI E., PELLERI F., TERRADURA M., FRATTEGIANI M., SAVINI P., GROHMANN F., MORI P., TERZUOLO P.G. & PIUSSI P. 2016. *Opportunities for coppice management at the landscape level: the Italian experience*. IForest – Biogeosciences and Forestry, 9(5): 775–782.

MANETTI M.C., GIANNINI T., CHIANUCCI F., CASULA A. & CUTINI A. 2013. *Cambiamenti strutturali ed ecologici in cedui di leccio in Sardegna a 25 anni dal taglio di avviamento ad alto-fusto*. Annals of Silvicultural Research, 37(1): 22–28.

MARZILIANO P.A. 2009. *Analisi quali-quantitativa della necromassa in cedui invecchiati di leccio (*Quercus ilex *L.) del Gargano*. Forest@, 6: 19–28.

MAYOR X. & RODÀ F. 1993. *Growth response of holm oak (*Quercus ilex *L) to commercial thinning in the Montseny mountains (NE Spain)*. Annales des sciences forestières, INRA/EDP Sciences, 50(3): 247–256.

MERCURIO R. 2020. *Pino domestico. Estetica e coltivazione*. Edizione digitale presso Youcanprint Self-Publishing. pp. 136.

MERCURIO R. 2021. *Boschi neri di leccio. Coltivazione, conservazione, restauro*. Edizione digitale presso Youcanprint Self-Publishing. pp. 119.

MERCURIO R., MUSARELLA C.M. & SPAMPINATO G. 2022. *Tipologie Forestali della Calabria*. Edizione digitale presso Youcanprint Self-Publishing. pp. 218.

MERCURIO R. & SPAMPINATO G. 2006. *I tipi forestali delle Serre Calabresi*. Edizioni Laruffa, Reggio Calabria. pp. 205. ISBN 88–7221–308–8

MONTOYA J.M.1989. *Encina y encinares*. Mundi Prensa. Madrid. pp. 133.

PANAIOTIS C., LOISEL R. & ROUX M. 1998. *Analyse de la réponse de la vegetation aux trouées naturelles dans une futaie agée de* Quercus ilex *L. en Corse (Ile méditerranéene)*. Canadian Journal of Forest Research, 28: 1125–1134.

PIGNATTI S. 1998. *I boschi d'Italia*. UTET, Torino. pp. 677.

PIUSSI P. & ALBERTI G. 2015. *Selvicoltura generale*. Compagnia delle Foreste. Arezzo, pp. 434.

PIZARRO-TOBÍAS P., FERNÁNDEZ M., NIQUI J.L., SOLANON J., DUQUE E., RAMOS J-L. & ROCA A. 2015. *Restoration of a Mediterranean forest after a fire: bioremediation and rhizoremediation field-scale trial.* Microbial Biotechnology, 8(1): 77–92.

PRATS S., MALVAR M., MARTINS M.A.S. & KEIZER J.J. 2014. *Post-fire soil erosion mitigation: a review of the last research and techniques developed in Portugal.* Cuadernos de Investigacion geografica, 40(2): 403–427.

PULIDO F.J. & DIAZ M. 2005. *Regeneration of a Mediterranean oak: a whole cycle approach.* Ecoscience, 12(1): 92–102.

PUXEDDU M. & DELOGU G. 2007. *Esperienze della Regione Sardegna nella valutazione e gestione delle aree forestali percorse dal fuoco. Un caso di studio: S. Niccolò Gerrei.* L'Italia Forestale e Montana, 62(1): 15–26.

QUEZEL P. & MEDAIL F. 2003. *Ecologie et biogéographie des forets du bassin méditerranéen.* Elsevier. Paris. pp. 571.

RETANA J., RIBA M., CASTEL C. & ESPELETA M. 1992. *Regeneration by sprouting of holm oak (*Quercus ilex*) stands exploited by selection thinning.* Vegetatio, 99–100: 355–364.

SARDIN T., ARGILLIER C. & BOURDENET P. 2003. *Amélioration des techniques de reconstitution en espècies feuillues de la forêt méditerranéenne.* Forêt Méditerranéenne, 24(4): 451–464.

SAVINI P. 2010. *Nuove tecniche d'intervento nei boschi cedui: l'esperienza del progetto Summacop.* In V. Carraro & T. Anfodillo (eds), *Atti del 46° Corso. Gestione multifunzionale e sostenibile dei boschi cedui: criticità e prospettive.* Università di Padova/Dip.TESAF. Pubblicazione del Corso di Cultura in Ecologia. pp. 73–84.

SCARFÒ F., MERCURIO R. & DEL PESO TARANCO C. 2011. *Assessing visual impacts of forest operations on a landscape in the Serre Regional Park of southern Italy.* Landscape and Ecological Engineering (LEE), 9(1): 1–10.

SEIGUE A. 1985. *La foret circummediterranéenne et ses problèmes.* Maisonneve et Larose. Paris. pp. 502.

SERRADA R., BRAVO A., SANCHEZ I., ALUE M.M., ELENA R. & SAN MIGUIEL A. 1998. *Conversion into high forest in coppices of* Quercus ilex *subsp* ballota *in central region of Iberian Peninsula.* Annali

Istituto Sperimentale per la Selvicoltura, 27 (1996): 149–160.

SERRADA HIERRO R., BRAVO J.A. & ROIG S. 2004. *Brotacion de encina (*Quercus ilex *subsp.*ballota*) con edades elevadas. Experiencias en el monte de Riofrio (Segovia)*. Investigacion Agraria Sistemas y Recursos Forestales, Fuera de Serie: 127–141.

SFERLAZZA S., MAETZKE F.G., IOVINO M., BAIAMONTE G., PALMERI V. & LA MELA VECA D.S. 2018. *Effects of traditional forest management on carbon storage in a Mediterranean holm oak (*Quercus ilex *L.) coppice*. iForest – Biogeosciences and Forestry, 11: 344–351.

SOLER M. & SALA M. 1992. *Effects of fire and of clearing in a Mediterranean* Quercus ilex *woodland: An experimental approach*. Catena, 19(3–4): 321–332.

SUSMEL L., VIOLA F. & BASSATO G. 1976. *Ecologia della lecceta del Supramonte di Orgosolo. III Contributo: produzione primaria, produzione secondaria (erbivori) condizioni attuali e possibilità di conservazione*. Annali del Centro di Economia Montana delle Venezie X (1969–70). pp. 261.

TERRADAS J. 1999. *Holm Oak and Holm Oak Forests An Introduction*. In F. Rodà, J. Retana, C.A. Gracia & J. Bellot (eds), *Ecology of Mediterranean Evergreen Oak Forests*. Springer. Berlin, Heidelberg. pp.3–14. https://doi.org/10.1007/978-3-642-58618-7_1.

TRABAUD L. 1996. *Résponse du chêne vert et du chêne blanc à l'action du feu*. Forêt Méditerranéenne, 17(3): 243–252.

VALLEJO V.R., ALLEN E.B., ARONSON J., PAUSAS J.G., CORTINA J. & GUTIÉRREZ J.R. 2012. Restoration of Mediterranean-type woodlands and shrublands. In J. van Andel & J. Aronson (eds), *Restoration Ecology: The New Frontier*, 2nd edn. Wiley-Blackwell Publishing. Oxford. pp.130–144.

VILLAR-SALVADOR P., NICOLAS PERAGON J.L., HEREDIA GUERRERO N. & USCOLA FERNANDEZ M. 2013. Quercus ilex *L*. In J. Pemán, Navarro-R.M. Cerrillo, J.L. Nicolás, M.A. Prada & R. Serrada (eds), *Producción y Manejo de Semillas y Plantas Forestales*, vol 2. Organismo Autónomo Parques Nacionales, Ministerio de Agricultura, Alimentación y Medio Ambiente. pp. 226–250.

CHAPTER 6

SILVICULTURE AND RESTORATION OF *QUERCUS SUBER* WOODLANDS

ROBERTO MERCURIO

6.1 Distribution range and types

Cork oak is a markedly "Atlantic" species widespread in the western Mediterranean basin. As a consequence of its peculiar ecology in Italy the species occurs especially in the Tyrrhenian sector (De Philippis, 1936) and in the two main islands, Sicily and in Sardinia (here primarily in the Alta Gallura, which indeed represents the Sardinian industrial district of cork). Along the Adriatic side *Quercus suber* is found in the province of Brindisi (Apulia), mostly in the Piana di Brindisi district, and more occasionally in coastal and subcoastal zones between the towns of Fasano and Ostuni and in some scattered stands of the Tavoliere Salentino plateau (Carella, 2008). The cork oak woods grow between 0 and 800 metres above sea level in Calabria (Mercurio et al., 2022), and 900 metres in Sicily and Sardinia, also if their optimum conditions are within 400 metres above the sea level. The cork oak woodlands are of great socio-economic importance and play a fundamental role in the Mediterranean landscape (Matvejevic, 1999; Camarda, 2005), in particular the anthropogenic landscapes show remarkable biological-cultural value (Montero et al., 2000; Olea et al., 2006; Serrada et al., 2008).

Fig. 6–1. Monumental cork oak. Alentejo, Portugal, 1992 (Photo Roberto Mercurio).

Fig. 6–2. Cork oak individuals in the Piana di Brindisi district, Apulia-South Italy (Photo Rocco Carella).

The classification of cork oak woodland types is not easy (Del Favero, 2008) since human management has increased the heterogeneity of the coenoses at local and regional level. Indeed, the cork oak has been widely spread for its economic importance, also for its connection with the wine production industry (Dettori & Filigheddu, 2016) and grazing. The cork oak has also been used in many other sectors in different parts of the Calabria region: to produce coal for manufacturing and industrial activities such as silk spinning, ashlar to produce liquorice, in the lime production process and above all in the steel industry. Other areas of use were domestic use for cooking and heating (Barreca et al., 2010), tannin production for leather (Caridi & Iovino, 2002), fishing net tanning and floats for fishing. Now cork is also used for footwear, packaging, insulation for the building industry and in handicrafts.

For these reasons, the European classification of forest types described for cork oak (Barbati et al., 2006, 2014) are still very general:

6.9.1 Mediterranean evergreen oak forest
Relationship to Annex I of "Habitat" Directive 92/43/EEC
Dehesas with evergreen *Quercus* spp. (code 6310)
Quercus suber forests (code 9330)

Relationship to CORINE Biotopes
45.21 – Tyrrhenian cork-oak forests (45.213 – Sardinian cork oak forests; 45.214 – Central Italian cork-oak forests; 45.215 – Southern Italian cork oak forests).

Relationship to EUNIS Habitat Classification
G2.11 – Mediterranean evergreen oak woodland.

In Italy two macro vegetation types have been described: xeric cork oak woodlands with holm oak, where other Mediterranean sclerophiylous species can be observed (Scarascia Mugnozza & Schirone, 1983; Mercurio & Spampinato, 2001, Bacchetta et al., 2009; Beccarisi et al., 2010); and mesophilous cork oak woodlands where *Quercus suber* is mixed with species of the group of pubescent oak (*Quercus pubescens, Quercus virgiliana*), *Quercus dalechampii* and Turkey oak (Mercurio & Spampinato, 2001; Quezel & Medail, 2003; Mercurio et al., 2009, 2022; Barreca et al., 2010; Mercurio, 2015).

Del Favero (2008) differentiates the following forest types for the southern Italian regions:

- cork oak woodlands of the meso-xeric soils with pubescent downy oak;
- cork oak woodlands of the xeric soils with Mediterranean sclerophyllous species;
- secondary cork oak woodlands (mainly addressed to cork stripping);
- cork oak woodlands of the volcanic soils.

The anthropogenic forest systems occurring in Italy show some similarities to those of other Mediterranean countries:

6.1.1 Silvosystems

Cork oak woodlands, pure or mixed with other broadleaved species and shrubs, where cork stripping can also be done (Tuscany, Latium, Calabria, Sicily). These are mostly high forests occasionally grazed, while coppicing is rare. Bernetti (1995) reports a case of coppice in Tuscany concerning a holm oak dominated coenoses with other sclerophyllous species and standards of cork oak. Holm oak–cork oak mixed coppices are common in Calabria (southern Italy), since abandonment, overgrazing, persistent fires and improper cork extractions facilitated *Quercus ilex* affirmation in cork oak woodlands (Mercurio et al., 2009; Barreca et al., 2010).

Fig. 6–3. Cork oak coenoses partially interested by forest utilization. Valle dell'Angitola, Calabria region, southern Italy (Photo Roberto Mercurio).

Fig. 6–4. Holm oak and cork oak mixed coppice. Calabria region, Thyrrhenian sector (Photo Roberto Mercurio).

6.1.2 Cork oak plantations

Management strategies for natural cork oak forests have focused in recent years on multifunctionality. For this reason, in Sardinia in recent years cork oak plantations are planned to ensure higher quality product, constant and quicker production processes, and finally to cope with increasing labour costs and a rapid growth in demand for job security (Abeltino et al., 2000; Dettori et al., 2001). The new models adopted since the 1990s consider the typical methods used in arboriculture: plantations limited to the optimal geopedological and morphological (inclines within 25%) stands, full deep ploughing and the use of good quality of selected nursery plant material. The model requires an initial stand density of 750–800 plants per hectare (3x4 metres) or 3 x 3 metres (AA.VV., 2002). At the same time, initiatives promote the value of cork through forest certification and supply chain traceability (Dettori et al., 2001, 2009; Pollastrini et al., 2018).

6.1.3 Agro-silvopastoral systems

Savanna-like systems without a shrub layer able to perform multiple functions are also frequent in Sardinia. The best-known examples are the Spanish *Dehesa* (south-west) and the Portuguese *Montado* (south-east) (Souto Cruz & Monteiro Alves, 1987; Pinto-Correia, 1993; Joffre et al., 1999; Eichhorn et al., 2006) whose origin dates back over 1,000 years ago (San Miguel Ayanz, 1994). The *Dehesa* is a traditional type of Mediterranean forest management where low density woodlands of native species (*Quercus ilex* – the most widespread – *Q. suber, Q. pyrenaica*, etc.) are interposed with crops (Gomez-Limon & de Lucio Fernandez, 1999). The degree of canopy cover varies between 10 and 60% (Serrada & San Miguel, 2008), but the optimal goes from 30 to 50% (Montero et al., 2000). A density tree cover layer as low as 60 trees per hectare (Pizzurro et al., 2010) has important effects on soil fertility and forage production in areas characterized by oligotrophic soils and dry summer period (Joffre et al., 1999; Gallardo, 2003; Moreno et al., 2007). The *Dehesa* is a multifunctional system, which can simultaneously support forest production (firewood, acorns, cork), wild breeding livestock, hunting and agricultural production, as well as honey production, mushrooms and aromatic herbs collection, all without irreversibly damaging its functionality (Olea & San Miguel-Ayanz, 2006; Linares, 2007; Serrada & San Miguel, 2008).

Fig. 6–5. Dehesa. *Extremadura, Spain (Photo Roberto Mercurio).*

Fig. 6–6. Cork oak plantations in spaced rows to allow grazing. Castel Porziano Presidential Estate, Latium region-Italy (Photo Roberto Mercurio).

6.1.4 Silvopastoral systems

Wooded pastures occurring in Sardinia, with cork oaks scattered in permanent meadows and pastures (Dettori & Filigheddu, 2004) are referred to as the considered system. In this system can also be listed the pastures of northern Sardinia, where cork production and forage are being pursued simultaneously. The system has a remarkable cultural and landscape value and, at the same time, is still nowadays a source of high income due to different revenues resulting from cork, meat and milk production. It also provides indirect fire control (Dettori et al., 2001; Del Favero, 2008). Despite this, Vacca (2000) highlighted how overgrazing, fire and soil tillage are responsible for severe soil degradation.

Fig. 6–7. Scattered cork oak in regenerated pastures. Sardinia (Photo Roberto Mercurio).

6.2 Elements of reproductive biology

6.2.1 Gamic regeneration

The cork oak has annual and biennial acorn maturation (Corti, 1953; Scarascia Mugnozza & Schirone, 1983, Díaz-Fernandez et al., 2004). Large seed production starts from the age of 15–20 years, with frequency of 3–4 years (Gambi, 1989; Monteleone et al., 2002). Seed dispersal occurs due to gravity (barochory) within the crown area, and zoochore, in this case reaching medium–long distances through rodents and birds (Pausas et al., 2009).

Fabbio (2009) reports acorn production that ranges from 15–30 kilograms up to 50–100 kilograms for each standard of cork oak.

Dissemination occurs in fall; part of the seed production is preyed on by birds and mammals, both in pre- and in post-dispersal. Laboratory results have certified that the germination capacity of seeds is 70–90% (Piotto et al., 2001); anyway, environmental factors play a key role. Seed germination depends on the interaction of many environmental conditions, while acorn size and plant cover are key factors determining seedling emergence and survival (Arosa et al., 2015).

The impact of shrubs on cork oak seedlings in the Mediterranean environment is frequently discussed to determine positive (protection) and negative (competition) impacts.

The natural regeneration of cork oak takes place easily in open fields (Di Pasquale & Garfi, 1998; Mercurio & Spampinato, 2001) but it can succumb to severe drought (Gambi, 1989). For this reason, according to Boudy (1954), Bosetto (1955) and Gambi (1989), cork oak regeneration takes advantage of maquis presence on its sides, which also ensures protection from animal bites. However, excessive density may compromise seedling growth. Pausas et al. (2006) observed the negative correlation between shrub vegetation and seedlings establishment. In the same way, Pons & Pausas (2006) have found that seedling density is frequently low under dense shrub cover, while Acacio et al. (2007) confirmed this in Portugal with specific reference to *Cistus* sp.

Some authors (Acacio et al., 2007; Plieninger et al., 2010) wisely concluded that shrub cover is favourable to seedlings survival as it mitigates the thermal stress typical for Mediterranean summers (Gómez-Aparicio et al., 2004).

Fig. 6–8. Cork oak seedlings under partial coverage of mother-tree (Photo Roberto Mercurio).

6.2.2 Agamic regeneration

Cork oak may resprout after a disturbance event, from stumps, roots, stems and also branches (this is unique to cork oak) (Pausas, 1997). Root suckers are a powerful method of diffusion for *Q. suber* L. and also for *Q. crenata* Lam., *Quercus suber* x *Quercus cerris* hybrid. The duration of the dormant buds in the cork oak and the consequent resprouting capacity is 120 years, even after fire (Gambi, 1988).

6.3 Cork oak woodlands silviculture

The silvicultural treatment of spontaneous cork oak woodlands for cork production involves a number of thinnings to select individuals with the best stem shape and cork quality; at the same time the considered interventions model the stand density. There are different points of view among authors about thinning intensity and therefore optimal stand density (Natividade, 1950; Bosetto, 1955; Montoya, 1988; Montero & Canellas, 1999, 2003; Montero & Lopez, 2008).

Age (years)	Density (trees per hectare)
20	700
30	400
100	200

Tab. 6–1. Silvicultural model by Bosetto (1955).

Age (years)	Girth (at 1.30 m)	Density (rees per hectare)
30	60	330
60	120	100
90	180	51

Tab. 6–2. Silvicultural model by Montoya (1988).

Before trees reach a bark circumference over 60 centimetres (the age of the first cork extraction) thinning is necessary to create more space between the individuals and to select the most productive one. The first thinning is carried out at 15–20 years and the second thinning (supported by a formation pruning up to 2–2.5 metres) at 25–35 years, one year before the first cork extraction.

Girth (at 1,30 m)	Density (trees per hectare)
60	450–500
70	350–450
80	300–350
90	250–300
100	200–250
110	175–200
120	150–175
130	125–150
140	100–125
150	95–100
160	90–95
170	85–90
180	80–85
190	75–80
200	70–75
>200	65–70

Tab. 6–3. Silvicultural model by Montero & Canellas (1999, 2003), Montero & Lopez (2008).

Natividade (1950) adopts a different criterion based on the assumption that mature cork oak individuals take advantage of the maximum solar irradiance of the crown; thus, the distance between the trees should be 2.5 times the crown mean radius.

Stand coefficient and spacing coefficient are also two important indicators. The first is the ratio between the total crowns cover (vertical projection on the ground) area and the surface of the cork oak woodland, which should be between 0.58 and 0.78 (Gambi, 1989). The second is that number which, multiplied by the diameter of the unitary mean crown projection area, gives the distance between the plants (with a distribution as if trees were in squares), and should be between 1.25 and 1.00 (Gambi, 1989).

The rotation length, although it becomes an uncertain concept for cork oak woodlands, varies from 90–110 to 140–150 years on the most fertile soils (Gambi, 1989).

Pruning plays an important role from the biological point of view (Natividade, 1950) and because the cork oak tends to ramify a lot. In the different stages of plant growth can be distinguished:

Formation pruning. This gives the plant the selected shape (straight stem) to enhance harvest operations (it is carried out up to a height of 2.5–3 metres, starting from the third year up to the first cork extraction). It consists mainly of the reshaping of the crown and in the elimination of low branches.

Yield pruning. If this is of moderate intensity it allows the shape of the crown and the vitality of the plant to be improved, therefore promoting the production of acorns and cork. It is carried out every 20–30 years, in the third winter following the cork stripping, reducing no more than 20% of the crown volume and leaving some branches in the middle of the crown that would overshadow first-order branches and assure a round-shaped crown.

Rejuvenation pruning. This is an extraordinary intervention targeted at diseased and damaged trees to stimulate recovery, which can include cutting up to 30–40% of the crown volume.

Pruning must not be carried out simultaneously with cork extraction: that is, pruning must be performed a few years before/after extraction. Even these are precautionary measures that are not scientifically proven. There have been contrasting results concerning the implication of pruning on the production of acorns and cork, but certainly an intensity over 35–40% of the canopy biomass causes negative effects (Montero & Curras, 1991). However, the effectiveness of pruning in terms of both quality and quantity is hard to prove because of the difficulty of assessing different situations (Cañellas & Montero, 2002).

Traditional silvicultural treatment, as a uniform shelterwood system, has been proposed with regard to even-aged monolayer stands (Bosetto, 1955; Bernetti, 1995; Montero et al., 2015; Pasalodos-Tato et al., 2018), even though there is no objective evidence in Italy. According to this, the seeding cut, coinciding with a year of cork stripping, involves 50% in number of trees; the overstorey removal cut is made after about ten years following one cork stripping. Rotation length is 80 years in coenoses exposed to drought (termo-xerophilous stands or with shallow soils) and 120–130 years in better conditions (less dry stands or most fertile soils).

For uneven-aged stands there is no proven application of selection systems. Indeed, though the single selection system has been proposed in Sardinia (Barneschi, 1984), at the moment there is no effective proof (Dettori et al., 2001). Montero & Canellas (2003) and Montero & Lopez (2008) emphasize that since cork oak is a light-demanding species it does not fit with typical uneven-aged structures with continuous seed regeneration where all age classes are represented, with possible exceptions in stands with two or more age classes or in open canopy stands. However, Vignes (1990), for cork oak irregular stands (*futaie jardinée* in French) in Var (France), indicates that cork extraction concerning about one-third of the total trees (where this is possible) and agamic regeneration through stump sprouts of the no longer productive individuals, is the reason why it cannot be considered a customary selection system.

More frequently a selection cut was applied without precise written rules, with fellings carried out exclusively on no longer productive trees of about 80–130 years; natural regeneration is assured by shoots resprouting from stumps and seedlings; also overgrazing can significantly affect regeneration.

In the same way in irregular stand structures modular cuts were proposed by Mercurio & Saba (1998), consisting of felling of no longer productive trees of about 130 years, with simultaneous localized clearing of shrubs to maintain a 60% coverage given by 200–300 large individuals.

Traditional cork stripping is a delicate manual process requiring skilled workers (*scorzini* in Italian slang), whose only tool is a small axe. Having cleaned the shrubs around the tree to facilitate access, they perform two circular horizontal incisions, one at the base of the tree and the other at the maximum harvesting height (established as a function of the stem circumference at 1.30 metres above the ground), and then one or two vertical incisions (connection cuts), without reaching and damaging the vascular cambium below (phellogen). Planks of cork are subsequently pulled out (separated) from the stem with the axe handle; these are therefore distinguished according to quality:

- virgin cork or first cork, used for agglomerates after trituration;
- second cork or gentle cork, used for stoppers and for agglomerates;
- extra quality cork, used for stoppers of high quality and disks.

Fig. 6–9. Cork planks piled after harvesting. Gallura, Sardinia (Photo Roberto Mercurio).

Mechanical cork harvesting is relatively recent and limited (Oliveira & Costa, 2012).

Stems and branches cannot be stripped before reaching a minimum perimeter: in Italy this is a circumference outside the bark of 60 centimetres at 1.30 metres in height, corresponding to an age of 25–30 (50–55) years. With the first cork stripping so-called virgin cork or first cork is obtained; the following ones, carried out every 9–10 years, provide the second cork or gentle cork and sometimes an extra quality cork. During its life a cork oak can bear 6–7 extractions in the worst stands and up to 9–10 in the better ones; after that, the trees undergo biological and productive decline and need to be replaced.

Graduate cork harvesting, due to non-simultaneous extraction in the management unit, enhances habitat heterogeneity and constitutes a recommended strategy. However, the consequent cork age heterogeneity of the management units causes an undesired dispersion of cork harvesting activities, with significant negative impacts on revenues (Costa et al., 2001; Oliveira & Costa, 2012).

In Italy, the maximum harvesting height at the time of the first cork extraction must not exceed two times the circumference outside the bark at breast height (CBH) and three times in the following ones, in any case no later than the circumference becoming less than 45 centimetres.

The cork harvesting coefficient (HC), defining the ratio between the maximum harvesting height and stem circumference at 1.30 metres above the ground (CBH), varies from 1.5 (less vigorous trees) to 2.5 (more vigorous trees).

Cork is harvested exclusively between mid-spring (mid-May) and summer (mid-August) when the phellogen is physiologically active and swollen, and at its maximum activity (Dettori et al., 2001); nevertheless, this period can vary in other Mediterranean and north African countries (Costa et al., 2002, 2003, Oliveira & Costa, 2012). In recent years, a review of stripping procedures has been taking place to mitigate plant stress (Pollastrini et al., 2018; Cutini et al., 2019).

During the production cycle in a good tree, 5–6 or 8–10 cork strippings are possible, providing 100–200 kilograms of raw cork. This means, annually, a production of 5–6 kilograms during the first cork stripping, and 20–30 kilograms when the trees are in full production (Gambi, 1989).

6.4 Restoration of degraded cork oak woodlands

The decline in cork oak woodlands is reported in many areas of the Mediterranean basin; it becomes manifest with loss of vegetative vigor of trees, higher sensitivity to physiological stress (e.g. drought) and subsequent pathogens attacks, the lack of natural regeneration with the consequent risk of cork oak disappearance (Belghazi et al., 2001; Pausas et al., 2006; Costa & Pereira, 2007; Costa et al., 2010; Acácio & Holmgren, 2014). Nevertheless, as cork oak has coevolved in the Mediterranean environment due to its ecological peculiarities, enabling the species to resist stress factors such as fire, many other complex causes are responsible for the cork oak decline (Montero et al., 2004; Costa et al., 2010). Among these, cork stripping remains the most important stress factor (Oliveira & Costa, 2012) affecting its biological decline. Cork oak woodlands are systems of high environmental and economic importance; thus, active restoration techniques have been developed throughout the Mediterranean basin (Caro, 1916; Andreuccetti, 1957; Vignes, 1990; Moreira et al., 2006; Aronson et al., 2009). The restoration of cork oak degraded stands requires significant initial investment, with no chance of economic return until 30–40 years. For this reason, public financial support should be provided for cork oak woodland restoration, given its ecosystem services satisfaction – such as biodiversity conservation and carbon storage (Ovando et al., 2009; Bugalho et al., 2011; Oubrahim et al., 2015). The purpose of cork oak woodlands restoration is to obtain diversified structures according to different stand conditions and dendrological composition, capable of ensuring higher resistance and resilience in the face of climate change and biological and human disturbances, and to meet at its best the needs of local peoples.

Every intervention should duly consider the biology, ecology and dynamics of the cork oak and the habitat in which it occurs (Acácio & Holmgren, 2014) focusing in particular on natural regeneration dynamics (Torres Alvarez, 2003). Furthermore, measures should also consider variables related to climate change that could trigger migration and local extinctions of the species, in particular in those districts exposed to the intensification of dry periods (Pereira et al., 2009; Costa et al., 2014, 2016; Verdinelli et al., 2017), and at the same time also new areas of potential expansion (Vessella & Schirone, 2013).

Fig. 6–10. Wide cork oak coenoses heavily degraded by recurrent fires and overgrazing. Calabria, southern Italy (Photo Roberto Mercurio).

Fig. 6–11. Degraded mesophilous coenoses of cork and mixed with Turkey oak. Calabria, southern Italy (Photo Roberto Mercurio).

The most important forest restoration activity is post-fire recovery, consisting of direct interventions to accelerate the reconstitution of the burned forest communities (Jacquet & Prodon, 2007).

Recurrent forest fires induce the gradual disappearance of cork oak, favouring shrubby vegetation such as *Erica* sp., *Cistus* sp., *Calicotome* sp. and *Myrtus communis* (Mercurio & Spampinato, 2001). On the other hand, in the absence of disturbance cork oak coenoses evolve towards holm oak forests (Di Benedetto et al.,1984; Pignatti, 1998), at least in the hottest and driest areas.

Cork oak is fire-resistant, being partially protected against fires by bark insulation (the cork) and its capacity to resprout (above-ground biomass recovery) after fire. Plants with virgin cork are consequently more fire-resistant than individuals already debarked (Barberis et al., 2003; Moreira et al., 2007). Tree vulnerability to fire is significantly higher in plants with thin bark (both young and recently debarked) and decreases until bark is about 4 centimetres thick (Catry et al., 2012). Cork oak is the only European forest species capable of resprouting from stem through epicormic buds located up on the plant (Pausas, 1997; Letreuch-Belarouci et al., 2009; Pausas et al., 2009; Catry et al., 2010; Acacio & Holmgren, 2014). Regarding crown recovery, the most important variables for its success are bark thickness, charring height and tree diameter (Catry et al., 2009).

A degradation index has been tested in the degraded oak woodlands of Sicily (Modica et al., 2015) to prioritize interventions during the restoration process.

6.4.1 Case study: restoration of cork oak stands through vegetative regeneration

Among the preliminary interventions, the extraction of flamed cork can be expected in the year following the fire passage when the plant has been already debarked for some years, if not seriously damaged. In the other cases, felling below the ground becomes more effective (Dettori et al., 2001; AA.VV., 2002).

Fig. 6–12. Debarking of burned cork oak individuals in Sardinia (Photo Roberto Mercurio).

It has been proved that timely coppicing after fire (before the following growing season) can reduce mortality and accelerate the rapid recovery of burned plants (Barberis et al., 2003; Pintus & Ruiu, 2004; Sirca et al., 2014). Other authors suggest waiting until the next spring for coppicing (Amo & Chacón, 2003), or the second autumn for a better evaluation of plants' vegetation conditions (Beltran, 2004). Felling below ground is effective to stimulate resprouting and reduce stump mortality (Milella et al., 1989; Pintus et al., 1995; Mercurio & Saba, 1998). Dettori et al. (1996) argue that the percentage of regrowth decreases progressively with more cork strippings. 5–8 years after the felling below the ground, stools are thinned releasing 1–2 per stump (Hasnaoui, 1991; Dettori et al., 2001).

Fig. 6–13. Stump lowering (final operation of the felling below the ground) with the typical axe, marra in Italian (Photo Roberto Mercurio).

6.4.2 Case study: restoration of cork oak stands through new plantations

In stands where the resprouting capacity is drastically reduced, seed dissemination is scarce, or where high disturbance level is recorded, Bosetto (1955) proposed a clear cut – either all at once or several times at intervals of five years. In this case regeneration must be artificial if there are no sprouts from the stumps.

The recovery of stand density (when crown coverage is less than 40%) is performed through partial soil preparation. Past experiments in small terraces showed higher seedling growth than in holes, due to the greater depth and water-retention capacity of soils obtained with the first preparation technique (Mercurio & Saba, 1998). The most used species is cork oak, but in very degraded sites other broadleaved species such as wild olive (*Olea europaea* var. *sylvestris*) or wild pear (*Pyrus amygdaliformis*) can also be planted – both relevant for feeding fauna and wood quality (Mercurio & Bagnato, 2008).

Fig. 6–14. Cork oak plantation on small terraces. Sardinia (Photo Roberto Mercurio).

On steep terrain, mulching and trasversal barriers are essential for reducing soil erosion (Bautista et al., 2004; Fernandez et al., 2011; Vallejo et al., 2012). As a consequence of recurrent fires, dense *Cistus* sp. shrublands hinder the recovery of cork oak woodlands (Pausas et al., 2006; Acácio et al., 2007). In these cases, the following operations have been suggested:

- partial shrub removal (strip clearings 3–5 metres wide);
- seeding or plantation of cork oak seedlings (Perez-Devesa et al., 2008); in this case besides *Cistus* sp. removal, acorn seeding, or seedling plantations (or planting shrubs which are *nurse plant* to facilitate seed germination) and cork oak establishment (Gómez-Aparicio et al., 2004; Acacio et al., 2007; Acácio & Holmgren, 2014);
- release of perch trees to attract birds, such as the Eurasian jay (*Garrulus glandarius*) that are able to spread cork oak seeds over a

long distance (Walker & Del Moral, 2009).

Small clearings can gradually reclose because of seed regeneration or root suckers. Rejuvenation pruning in mature trees can be useful to stimulate crown revegetation, with the removal of up to 40% of the crown (withered, damaged or dead branches).

Within the Natura 2000 sites where the Habitat code 9330 (*Quercus suber* forest) is listed, an unsolved question is at present much discussed in Italy: that is, is it appropriate to promote the natural evolution of the cork oak coenoses avoiding all the silvicultural operations (Angelini et al., 2016), or maintain cork oak stands and consequently their economic, landscape and cultural value?

References

AA.VV. 2002. *Sughericoltura*. ERSAT. Cagliari: 202.

ABELTINO P., D'ANGELO M., BARBERIS A., DETTORI S., FALQUI A., FILIGHEDDU M.R., MANCHINU M. & MULAS G.B. 2000. *Risultati preliminari sulla risposta della quercia da sughero all'intensificazione colturale in Sardegna*. Atti II SISEF, pp. 89–93.

ACÁCIO V. & HOLMGREN M. 2014. *Pathways for resilience in Mediterranean cork oak land use systems*. Annals of Forest Science, 71: 5–13.

ACÁCIO V., HOLMGREN M., JANSEN P.A. & SCHROTTER O. 2007. *Multiple recruitment limitation causes arrested succession in Mediterranean cork oak systems*. Ecosystems, 10: 1220–1230.

AMO E., CHACÓN C. 2003. Recomendaciones selvícolas para alcornocales afectados por el fuego. Cuadernos Forestales, IPROCOR, Mérida.

ANDREUCCETTI E. 1957. *Origini, evoluzione, struttura e ricostituzione dell'antico "Bosco di S. Pietro" nel territorio di Caltagirone (Sicilia)*. Monti e Boschi, 8(8): 210–219.

ANGELINI P., CASELLA L., GRIGNETTI A. & GENOVESI P. (eds). 2016. *Manuali per il monitoraggio di specie e habitat di interesse comunitario (Direttiva 92/43/CEE) in Italia: habitat*. ISPRA, Serie Manuali e linee guida, 142/2016.

ARONSON J., PEREIRA J.S. & PAUSAS J.G. 2009. *Cork Oak Woodlands on the Edge Ecology, Adaptive Management, and Restoration*. Island Press. Washington DC, pp. 352.

AROSA M.L., CEIA R.S. COSTA S.R. & FREITAS H. 2015. *Factors affecting cork oak (*Quercus suber*) regeneration: acorn sowing success and seedling survival under field conditions*. Plant Ecology & Diversity. http://dx.doi.org/10.1080/17550874.2015.1051154.

BACCHETTA G., BAGELLA S., BIONDI E., FARRIS E., FILIGHEDDU R. & MOSSA L. 2009. *Vegetazione forestale e serie di vegetazione della Sardegna (con rappresentazione cartografica alla scala 1:350.000)*. Fitosociologia, 46(1) suppl. 1: 3–82.

BARBATI A., CORONA P. & MARCHETTI M. 2006. *European forest types. Categories and types for sustainable forest management and reporting*. European Environment Agency, EEA Technical report No. 9/2006.

BARBATI A., MARCHETTI M., CHIRICI G. & CORONA P. 2014. *European Forest Type and Forest Europe SFM indicators: tools for monitoring progress on forest biodiversity conservation*. Forest Ecology and Management, 321: 145–157.

BARBERIS A., DETTORI S. & FILIGHEDDU M.R. 2003. *Management problems in Mediterranean cork oak forests: post-fire recovery*. Journal of Arid Environments, 54: 565–569.

BARNESCHI L. 1984. *Modelli forestali della selvicoltura mediterranea*. Bollettino degli interessi sardi, Quaderno n.14, pp.73–87.

BARRECA L., MARZILIANO P.A., MENGUZZATO G., SCUDERI A. 2010. *Analisi strutturale e caratterizzazione della necromassa in sugherete della Calabria*. Forest@, 7: 158–168.

BARTOLINI G., BUSSOTTI F., ASENSIO A., GARCÍA C., CHESSA I., CICCARESE C., CICCARESE L., CROSTI R., CULLUM F.J., DI NOI A., GARCÍA-FAYOS P., LAMBARDI M., LISC MI., LUCCI S., MELINI S., MUÑOZ EINOSO J.C., MURRANCA S., NIEDDU G., PACINI E., PAGNI G., PATUMI M., PÉREZ GARCÍA F., PICCINI C., ROSSETTO C. & TRANNE G. 2001. *Schede informative sulla propagazione per seme degli alberi e degli arbusti più diffusi della flora mediterranea*. In B. Piotto & A. Di Noi (eds), *Propagazione per seme di alberi e arbusti della flora mediterranea*. ANPA. Roma, pp. 143.

BAUTISTA S., ROBICHAUD P.R. & BLADE C. 2009. *Post-fire mulching*. In A. Cerdà & P.R. Robichaud (eds), *Fire Effects on Soils and Restoration Strategies*. Science Publishers. Enfield, pp. 353–372.

BECCARISI L., BIONDI E., CASAVECCHIA S., ERNANDES P., MEDAGLI P. & ZUCCARELLO V. 2010. *La quercia da sughero (*Quercus suber *L.) nel Salento: analisi diacronica e sinfitosociologica (Adriatico meridionale, Italia)*. Fitosociologia, 47(2): 3–16.

BELGHAZI, B., M. EZZAHIRI, M. AMHAJAR & M. BENZIANE. 2001. *Régénération artificielle du chêne liège dans la forêt de la Maamora (Maroc)*. Forêt Méditerranéenne, XXII(3):253–261.

BELTRAN S.R. 2004. *Recommendations sylvicoles pou le suberaies affecté par le feu. Colloque Vivexpo 2004: le chène liège face au feu*. Vives. France, pp. 27.

BOSETTO G.1955. *Conservazione e miglioramento delle sugherete in Italia*. Atti del I Congresso Nazionale di Selvicoltura. Firenze, vol. I, pp. 1–48.

BOUDY P. 1950. *Économie forestière Nord-Africaine*, Tome II, Fascicule I, Monographies et traitements des espèces forestières: le chêne-liège. Larose. Paris, pp. 29–249.

BUGALHO M.N., CALDEIRA M.C., PEREIRA J.S., ARONSON J., PAUSAS J.G. 2011. *Mediterranean cork oak savannas require human use to sustain biodiversity and ecosystem services*. Frontiers in Ecology and the Environment, 9: 278–286.

CAMARDA I. 2005. *Il paesaggio della quercia da sughero (*Quercus suber *L.) nel Mediterraneo*. In C. Blasi & A. Paolella (eds), *Atti del III Congresso IAED Roma 4–6 dicembre 2003 Identificazione e cambiamenti nel paesaggio contemporaneo*. Papageno. Palermo, pp. 275–284.

CAÑELLAS I. & MONTERO G. 2002. *The influence of cork oak pruning on the yield and growth of cork.* Annals of Forest Science Springer Verlag/EDP Sciences, 2002, 59 (7), pp.753–760.

CARELLA R. 2008. *Le stazioni di* Quercus suber *del Brindisino*. Umanesimo della Pietra, Riflessioni 30: 159–166.

CARIDI D. & IOVINO F. 2002. *La presenza della quercia da sughero (*Quercus suber *L.)* in Calabria. L'Italia Forestale e Montana, 57(6): 513–532.

CARO E. 1916. *Restauración de alcornocales*. Revista Montes, 954: 684–690; 721–728.

CATRY F.X., MOREIRA F., DUARTE I. & ACÁCIO V. 2009. *Factors affecting post-fire crown regeneration of cork oak (*Quercus suber*) trees*. European Journal of Forest Research, 128: 231–240.

CATRY F.X., MOREIRA F., PAUSAS J.G., FERNANDES P.M., REGO F., CARDILLO E. & CURT T. 2012. *Cork Oak Vulnerability to Fire: The Role of Bark Harvesting, Tree Characteristics and Abiotic Factors*. PLoS ONE 7(6): e39810. doi:10.1371/journal.pone.0039810.

CATRY F.X., REGO F., MOREIRA F., FERNANDES P.M. & PAUSAS J. 2010. *Post-fire tree mortality in mixed forests of central Portugal*. Forest Ecology and Management, 260: 1184–1192.

CORTI R. 1953. *Sughere a maturazione biennale della ghianda in Puglia*. Nuovo Giornale Botanico Italiano n. s., 60 (4): 686–688.

COSTA A., BARBOSA I., ROUSSADOD C., GRAC J. & SPIECKERF H. 2016. *Climate response of cork growth in the Mediterranean oak (Q*uercus suber *L.) woodlands of southwestern Portugal*. Dendrochronologia, 38: 72–81.

COSTA A., MADEIRA M. & PLIENINGER T. 2014. *Cork oak woodlands patchiness: A signature of imminent deforestation?* Applied Geography, 54: 18 – 26.

COSTA A. & PEREIRA H. 2007. *Montados e sobreirais: uma espécie, duas perspectivas*. In J. Sande Silva (ed.), *Os Montados. Muito para além das árvores*. Árvores e Florestas de Portugal 3, 1st edn. Público, Comunicação Social SA & Fundação Luso-Americana para o Desenvolvimento. Lisboa: 17–37.

COSTA, A., PEREIRA, H. & MADEIRA, M., 2010. *Analysis of spatial patterns of oak decline in cork oak woodlands in Mediterranean conditions*. Annals of Forest Science 6: 204.doi:10.1051/forest/2009097.

COSTA A., PEREIRA H. & OLIVEIRA A. 2003. *Variability of radial growth in cork-oak adult trees under cork production*. Forest Ecology and Management, 175: 239–246.

COSTA A., PEREIRA H. & OLIVEIRA A. 2002. *Influence of climate on the seasonality of radial growth of cork oak during a cork production cycle*. Annals of Forest Science, 59: 429–437.

COSTA A., PEREIRA H. & OLIVEIRA A. 2001. *A dendroclimatological approach to diameter growth in adult cork-oak trees under production.* Trees, 15: 438–443.

CUTINI A., MUSCAS F., CARTA V., CASULA A., DETTORI S., FILIGHEDDU M.R., MALTONI S., PIGNATTI G. & ROMANO R., 2019. *Analisi e proposte per la valorizzazione della sughericoltura e della filiera sughericola Italiana.* Rete Rurale Nazionale, Consiglio per la ricerca in agricoltura e l'analisi dell'economia agraria. Roma.

DE PHILIPPIS A. 1936. *La sughera (*Quercus suber*) e il leccio (*Q.ilex*) nella vegetazione arborea mediterranea. Saggio di fitogeografia ed ecologia comparate.* Bulletin de la Silva Mediterranea Décembre 1935, pp.9–109.

DEL FAVERO R. 2008. *I boschi delle regioni meridionali e insulari d'Italia.* Cleup. Padova, pp. 466.

DETTORI S., FALQUI A. & FILIGHEDDU M.R. 1996. *Esperienze di recupero di sugherete percorse dal fuoco.* 2nd International Symposium on Cork, 12–14 September 1996, Pavia, pp. 129–132.

DETTORI S. & FILIGHEDDU M.R. 2016. *Il sughero in enologia. Analisi della filiera nazionale.* L'Italia Forestale e Montana, 71(6): 331–343.

DETTORI S. & FILIGHEDDU M.R. 2004. *Considerazioni sul ruolo delle foreste di sughera in aree a gestione naturalistica.* Italus Hortus, 4: 243–245.

DETTORI S., FILIGHEDDU M.R. & GUTIERREZ M. 2001. *La coltivazione della quercia da sughero.* DESE, Università di Sassari, pp. 138.

DETTORI S., MARONE E. & PORTOGHESI L. 2009. *Filiera delle produzioni forestali non legnose: produzione e raccolta tra sostenibilità e tracciabilità.* In O. Ciancio (ed.), *Atti del Terzo Congresso Nazionale di Selvicoltura per il Miglioramento e la conservazione dei boschi italiani.* Taormina (ME) 16–19 ottobre 2008. Accademia Italiana di Scienze Forestali II: 742–751.

DIAZ-FERNANDEZ P.M., CLIMENT J. & GIL L. 2004. *Biennial acorn maturation and its relationship with flowering phenology in Iberian populations of* Quercus suber. Trees, 18: 615–621.

DI BENEDETTO L., MAUGERI G. & POLI MARCHESE E. 1984. *Principali tappe del dinamismo della vegetazione delle sugherete della Sicilia sud orientale.* Notiziario fitosociologico, 19(2): 5–12.

Di Pasquale G. & Garfì G. 1998. *Analyse comparée de l'évolution del la régéneration de* Quercus suber *et* Quercus pubescens *après élimination du paturage en foret de Pisano (Sicilie-sud-orientale)*. Ecologia mediterranea, 24: 15–26.

Eichhorn M.P., Paris P., Herzog F., Incoll L.D., Liagre F., Mantzanas K., Mayus M., Moreno G., Papanastasis V.P., Plibeam D.J., Pisanelli A. & Dupraz C. 2006. *Silvoarable agriculture in Europe-past, present and future prospects*. Agroforestry Systems 67: 29–50.

Fabbio G. 2009. *Struttura coltivazione condizione e produttività del bosco all'attualità. In Fabbio G. (a cura di) Il bosco e l'allevamento della razza Cinta Senese*. ARSIA. Regione Toscana: 107–120.

Fernández C., Vega J.A., Jiménez E. & Fonturbel T. 2011. *Effectiveness of three post-fire treatments at reducing soil erosion in Galicia (NW Spain)*. International Journal of Wildland Fire, 20(1): 104–114.

Gallardo A. *2003. Effect of tree canopy on the spatial distribution of soil nutrients in a Mediterranean Dehesa*. Pedobiologia, 47: 117–125.

Gambi G. 1989. *Le sugherete II*. Monti e Boschi (Le schede), 40(2): I–XII.

Gambi G. 1988. *Le sugherete I*. Monti e Boschi (Le schede), 39(2): I–VIII.

Gómez-Aparicio L., Zamora R., Gómez J.M., Hódar J.A., Castro J. & Baraza E. 2004. *Applying plant facilitation to forest restoration in Mediterranean ecosystems: a meta-analysis of the use of shrubs as nurse plants*. Ecological Applications, 14: 1128–1138.

Gomez-Limon J. & de Lucio Fernandez J.V. 1999. *Changes in use and landscape preferences on the agricultural-livestock landscapes of central Iberian Peninsula (Madrid, Spain)*. Landscape and Urban Planning, 44: 165–175.

Hasnaoui B. 1991. *Régéneration naturelle par rejets de souche et par drageonnement d'une suberaie dans le Nord Ouest de la Tunisie*. Ecologia Mediterranea, 27: 79–87.

Jacquet K. & Prodon R. 2007. *Résilience comparée des peuplements de chene vert et de chene liège après incendies*. Revue Forestière Française, 59(1): 31–44.

Joffre R., Rambal S. & Ratte J.P. 1999. *The Dehesa systems of southern Spain and Portugal as a natural ecosystem mimic*. Agroforestry

Systems, 45: 57–79.

LETREUCH-BELAROUCI A.M., LETREUCH-BELAROUCI N., BENABDELI K. & MEDJAHDI B. 2009. *Impact des incendies sur la structure des peuplements de chène liège et sur le liège: le cas de la suberaie de Tlemcen (Algerie)*. Forêt Méditerranénne, 30(3): 231–238.

LINARES A.M. 2007. *Forest planning and traditional knowledge in collective woodlands of Spain: the Dehesa system*. Forest Ecology and Management, 249: 71–79.

MATVEJEVIC P. 1999. *Mediterranean. A cultural landscape*. University of California Press. Berkeley, Los Angeles, CA.

MERCURIO R. 2015. *Distribution et écologie des peuplements internes de chêne liège en Toscane (Italie)*. Forêt Méditerranéenne, XXXVI(1): 17–22.

MERCURIO R. & BAGNATO S. 2008. *L'oleastro: specie per la ricostituzione forestale nel Bacino del Mediterraneo.* Linea Ecologica, XL(1): 18–22.

MERCURIO R., BAGNATO S., FALCONE A., ROMANO G., SCARFÒ F. & SPAMPINATO G., 2009. *I tipi forestali della Comunità Montana dei Monti Reventino-Tiriolo-Mancuso.* Edizioni Laruffa. Reggio Calabria, pp. 98 + 10..

MERCURIO R., MUSARELLA C.M. & SPAMPINATO G. 2022. *Tipologie Forestali della Calabria*. Edizione digitale presso Youcanprint Self-Publishing, pp. 218.

MERCURIO R. & SABA G. 1998. *Expériences sur la réconstitution de la subéraie en Sardaigne (Italie)*. Annales de l'INRGREF n. sp., pp. 35–41.

MERCURIO R. & SPAMPINATO G. 2001. *Primo contributo alla definizione tipologica delle sugherete in Calabria.* Atti del III Congresso Nazionale SISEF 3: 483–490.

MILELLA A., DETTORI S., NUDDA G. & PODDIGHE D. 1989. *Differenti modalità di ricostituzione delle sugherete percorse da incendio*. Cellulosa e Carta 50 (1): 32–35.

MODICA G., MERLINO A., SOLANO F. & MERCURIO R. 2015. *An index for the assessment of degraded Mediterranean forest ecosystems*. Forest Systems, 24 (3): 1–13.

MONTELEONE I., GROOT S. PC., TESNIER K. & BELLETTI P. 2002. *Problematiche sementiere e vivaistiche in quercia da sughero*. Monti e

Boschi, 53(1): 41–45.

MONTERO G. & CAÑELLAS I. 2003. *Selvicultura de los alcornocales en Espana*. Silva Lusitanica, 11(1) 1:19.

MONTERO G. & CAÑELLAS I. 1999. *Manual de reforestacion y cultivo de alcornoque (*Quercus suber *L.)*. Mapa. Madrid: 103.

MONTERO G. & CURRAS R., 1991. *La poda del alcornocal (*Quercus suber *L.). Cuantificación de sus productos*. Ministerio de Agricultura, Servicio de Extensión Agraria, HD num. 18–19/90.

MONTERO G. & LOPEZ E. 2008. *Selvicultura de* Quercus suber. In R. Serrada, G. Montero, J.A. Reque (eds), *Compendio de Selvicultura Aplicada en España*. INIA. Madrid, pp.779–829.

MONTERO M.J., OBRADOR J.J., CUBERA E. & MORENO G. 2004. *The role of Dehesa land use on tree water status in Central-Western Spain*. In S. Schnabel and A. Ferreira (eds), *Sustainability of agrosilvopastoral systems – Dehesas, Montados*. Advances in GeoEcollgy 37, Catena Verlag, Reiskirchen, pp. 125–136.

MONTERO G., PASALODOS-TATO M., LÓPEZ SENESPLEDA E., RUIZ-PEINADO R., BRAVO A. & MADRIGAL G. 2015. *Modelos de selvicultura y producción de madera, frutos y fijación de carbono de los sistemas forestales de Andalucía*, vol. 1.1. In P. Campos and L. Díaz-Balteiro (eds), *Memorias científicas de RECAMAN*. Economía y selviculturas de los montes de Andalucía, Madrid: Consejo Superior de Investigaciones Científicas.

MONTERO G., SAN MIGUEL A. & CAÑELLAS I. 2000. *Systems of Mediterranean Silviculture "La Dehesa"*. Grafistaff S.L. Madrid.

MONTOYA OLIVER J.M. 1988. *Los alcornocales*. MAPA. Madrid, pp. 267.

MOREIRA F., DUARTE L., CATRY F. & ACÁCIO V. 2007. *Cork extraction as a key factor determining post-fire cork oak survival in a mountain region of southern Portugal*. Forest Ecology and Management, 253: 30–37.

MOREIRA F., QUEIROZ A.I. & ARONSON J. 2006. *Restoration principles applied to cultural landscapes*. Journal of Natural Conservation, 14: 217–224.

MORENO G., OBRADOR J.J. & GARCIA A. 2007. *Impact of evergreen oaks on soil fertility and crop production in intercropped Dehesas*. Agriculture, Ecosystems and Environment, 119: 270–280.

NATIVIDADE J.V. 1950. *Subericultura. Ministerio da Agricoltura, Pescas e Alimentaçao*. Direçcao Geral das Florestas. Lisboa, pp.387.

OLEA L. & SAN MIGUEL-AYANZ A. 2006. *The Spanish dehesa. A traditional Mediterranean silvopastoral system linking production and nature conservation*. 21st General Meeting of the European Grassland Federation. Badajoz (Spain). April 2006 Opening Paper, pp.1–15.

OLIVEIRA G. & COSTA A. 2012. *How resilient is* Quercus suber *L. to cork harvesting? A review and identification of knowledge gaps*. Forest Ecology and Management, 270: 257–272.

OUBRAHIM H., BOULMANE M., BAKKER M.R., AUGUSTO L. & HALIM M. 2015. *Carbon storage in degraded cork oak (*Quercus suber*) forests on flat lowlands in Morocco*. iForest, 9: 125–137.

OVANDO P., CAMPOS P., OVIEDO J.L. & MONTERO G. 2009. *Cost-benefit analysis of cork oak woodland afforestation and facilitated natural regeneration in Spain*. In J. Aronson, J.S. Pereira & J.G. Pausas (eds), *Cork oak woodlands on the edge. Ecology, adaptive management, and restoration*, 1st edn. Society for Ecological Restoration International, Island Press. Washington DC, pp 177–188.

PASALODOS-TATO M., PUKKALA T., CAÑELLAS I. & SÁNCHEZ-GONZÁLEZ M. 2018. *Optimizing the debarking and cutting schedule of cork oak stands*. Annals of Forest Science 75: 61 https://doi.org/10.1007/s13595–018–0732–8.

PAUSAS J.G. 1997. *Resprouting of* Quercus suber *in NE Spain after fire*. Journal of Vegetation Science, 8: 703–706.

PAUSAS J.G., MARAÑÓN T., CALDEIRA M. & PONS J. 2009. *Natural regeneration*. In J. Aronson, J. S. Pereira, J.G. Pausas (eds), *Cork oak woodlands on the edge. Ecology, adaptive management, and restoration*, 1st edn. Society for Ecological Restoration International, Island Press. Washington DC, pp. 115–124.

PAUSAS J.G., RIBEIRO E., DIAS S.G., PONS J. & BESELER C. 2006. *Regeneration of a marginal Quercus suber forest in the eastern Iberian Peninsula*. Journal of Vegetation Sciences, 17: 729–738.

PEREIRA J.S., CORREIA A.V. & JOFFRE R. 2009. *Facing climate change*. In J. Aronson, J. S. Pereira, J.G. Pausas (eds), *Cork oak woodlands on the edge. Ecology, adaptive management, and restoration*, 1st edn. Society for Ecological Restoration International, Island Press. Washington DC, pp. 219–226.

PEREZ-DEVESA M., CORTINA J.J., VILAGROSA A. & VALLEJO R. 2008. *Shrubland management to promote* Quercus suber *L. establishment*. Forest Ecology and Management, 255: 374–382.

PIGNATTI S. 1998. *I boschi d'Italia*. UTET. Torino, pp. 677.

PINTO-CORREIA T. 1993. *Threatened landscape in Alentejo, Portugal: the "Montado" and other "agro-silvo-pastoral" systems*. Landscape and Urban Planning, 24(1–4): 43–48.

PINTUS A., PAMPIRO F. & RUIU P.A. 1995. *Analisi di differenti interventi selvicolturali per il recupero di una sughereta percorsa da incendio*. Atti del Convegno "Arboricoltura da legno e politiche comunitarie" Tempio Pausania (SS) 22–23 giugno 1993, pp. 287–295.

PINTUS A. & RUIU P. 2004. *La réhabilitation des suberaies incendiées*. Colloques Internationaux Vivexpo – Le chêne-liège face au feu.

PIZZURRO G., MAETZKE F. & LA MELA VECA D. 2010. *Differences of raw cork quality in productive cork oak woods in Sicily in relation to stand density*. Forest Ecology and Management, 260: 923–929.

PLIENINGER T., ROLO V. & MORENO G. 2010. *Large-scale patterns of* Quercus ilex, Quercus suber, *and* Quercus pyrenaica *regeneration in Central-Western Spain*. Ecosystems, 13: 644–660.

POLLASTRINI M., CHIAVETTA U., CUTINI A., CASULA A., MALTONI S., DETTORI S. & CORONA P. 2018. *Indicators for the assessment and certification of cork oak management sustainability in Italy*. iForest, 11: 668–674.

PONS J. & PAUSAS J.G. 2006. *Oak regeneration in heterogeneous landscapes: the case of fragmented* Quercus suber *forests in the eastern Iberian Peninsula*. Forest Ecology and Management, 231: 196–204.

QUEZEL P. & MEDAIL F. 2003. *Ecologie et biogéographie des forets du bassin méditerranéen*. Elsevier. Paris, pp. 571.

SAN MIGUEL AYANZ S. 1994. *La Dehesa Espanola. Origen, tipologia, caracteristicas y gestion*. E.T.S. Ingenieros de Montes de Madrid-Fundacion Conde del Valle Salazar. Madrid.

SCARASCIA MUGNOZZA G. & SCHIRONE B. 1983. *Un bosco di sughera presso Brindisi*. Monti e Boschi, 34(6): 47–52.

SERRADA R. & SAN MIGUEL A. 2008. *Selvicultura en Dehesa*. In R. Serrada, G. Montero, J.A. Reque (eds), *Compendio de Selvicultura Aplicada en España*. INIA. Madrid, pp. 861–876.

SIRCA C., FILIGHEDDU M.R., ZUCCA G.M., CILLARA M., BACCIU A., BOSU S. & DETTORI S. 2014. *Long term researches on post fire recovery techniques of cork oak stands*. Proceedings of the Second International Congress of Silviculture, Florence, 26–29 November 2014, pp.491–496.

SOUTO CRUZ C. & MONTEIRO ALVES A.A. 1987. *Ecological fires influences on* Quercus suber *forest ecosystems*. Ecologia Mediterranea, XIII(4): 69–78.

TORRES ÁLVAREZ E. 2003. *Experiencias sobre regeneracion natural de alcornoque (*Quercus suber *L.)*. Cuad. Soc. Esp. Cien. For. 15: 37–47 (2003) Actas de la III Reunión sobre Regeneración Natural-IV Reunión sobre Ordenación de Montes.

VACCA A. 2000. *Effect of land use on forest floor and soil of a* Quercus suber *L. forest in Gallura (Sardinia, Italy)*. Land Degradation and Development, 11(2): 167–180.

VALLEJO V.R., ALLEN E.B., ARONSON J., PAUSAS J.G., CORTINA J. & GUTIÉRREZ J.R. 2012. Restoration of Mediterranean-type woodlands and shrublands. In J. van Andel & J. Aronson (eds), *Restoration Ecology: The New Frontier*, 2nd edn. Wiley-Blackwell Publishing, Hoboken, New Jersey, pp.130–144.

VERDINELLI M., YAKHLEF S.E.B., COSSU C.S., PILIA O. & MANNU R. 2017. *Variability of ant community composition in cork oak woodlands across the Mediterranean region: implications for forest management*. iForest, 10: 707–714.

VESSELLA F. & SCHIRONE B. 2013. *Predicting potential distribution of* Quercus suber *in Italy based on ecological niche models: Conservation insights and reforestation involvements*. Forest Ecology and Management, 304: 150–161.

VIGNES E. 1990. *Sylviculture des suberaies varoise*. Forêt Méditerranéenne, XII(2):125–130.

WALKER L.R. & DEL MORAL R. 2009. *Transition dynamics in succession: implications for rates, trajectories, and restoration.* In R.J. Hobbs & K.N. Suding (eds), *New models for ecosystem dynamics and restoration*, 1st edn. Society for Ecological Restoration International, Island Press. Washington DC, pp 33–49.

ACKNOWLEDGEMENTS

I would like to thank Professor Roberto Mercurio, Professor Roubina Ghattas, Professor Munir Otzurk and Professor Volkan Altay for their contributions.